THE LONG, LONG L

THE LONG, LONG LIFE OF TREES
Fiona Stafford

YALE UNIVERSITY PRESS
NEW HAVEN AND LONDON

For information about this and other Yale University Press publications, please contact:
U.S. Office: sales.press@yale.edu   yalebooks.com
Europe Office: sales@yaleup.co.uk   yalebooks.co.uk

Typeset in Adobe Garamond Pro and Copperplate by IDSUK (DataConnection) Ltd
Printed in Great Britain by TJ International, Padstow, Cornwall

Library of Congress Control Number: 2016944967

ISBN 978-0-300-20733-0 (cloth)

ISBN 978-0-300-22820-5 (pbk)

A catalogue record for this book is available from the British Library.

10 9 8 7 6 5 4 3 2

CONTENTS

Buds, Bark and a Golden Bough

THERE is a pine cone on my desk, about the size of a sparrow. It is too fat for my fingers to encompass fully, but I like to feel the rough, woody flakes in the warmth of my palm. When submerged in water, each smoothly curved scale fits as tightly as tortoiseshell and then, as it dries out, the tapered cone quietly relaxes into a tough, dry ball. As the gaps widen, the unyielding chocolate-brown chips begin to reveal caramel-coloured chevrons, which mark their more contracted form. I picked it up on holiday in Croatia three years ago and so, as it opens, each wooden spatula offers a small scoop of unremembered time – a sweltering trek through olive groves, an amphitheatre overlooking a busy harbour, a black octopus darting out from under a rock, creating drama in a quiet cove ringed by bright umbrellas and dark, stone pines.

Next to the cone is a little branch with a bunch of dried oak leaves, still firmly attached. There must be about forty, each one different in length, colour and curl. Their undersides are mostly like pale brown paper, lined with raised veins and marked with sporadic specks and flecks, but the topsides retain the deeper tan of polished, unstained leather. Their haphazard, waving, asymmetric outlines seem quietly anarchic. One reminds me of the last attempt at a pancake, when the batter is running out and no amount of tipping and turning will quite make a proper circle in the pan. The thin layers of crisped leaves harbour the smell of autumn and, if given a quick flick, will begin to mimic the scratching sound of the wind. The branch came from a mature oak a couple of miles from the house. I took it home when the land changed hands and the new owner began clearing old hedges and filling in ponds. Some of the acorns went into pots, some directly into odd corners of the garden to see if they would germinate. So far, the tree has been spared, but a few of its acorns have also sprouted into miniature versions with four or five small leaves of their own.

There are other things to plant. Black walnuts from a friend's tree, which are sitting side by side like basking toads, some already dry and smooth, others much darker and slightly sticky; all giving off varied hollow tones when tapped. I have no idea whether any of them will ever grow into trees. Their smell is more pungent than that of the oak leaves, a more insistent reminder of the world outside. And there is a horse chestnut gathered one September from the stately grounds of Chatsworth House. It should have been planted years ago, but now that it has hardened and has lost all its gloss I leave it to take its place with the other stolen mementos and promises. It sits beside a sliver of birch bark, half unrolled like a small vellum scroll or an unfilled cigarette paper. The whole house is full of things that were once alive, from the bead necklace and the bamboo bookcase to the oak floorboards and the olive wood fruit bowl, the pine chest and the cedar pencil, the beech bread bin and the bentwood chairs. Somehow the little thefts from trees I have seen for myself provide quicker connections to the natural environment.

The oak branch is my golden bough, offering immediate safe conduct from one world to another. It transports me to a particular day and tree, and then on to other oaks and their places, some of these known personally, others vicariously through things I have been told, or through poems and stories, photographs and paintings. Sometimes it will take me full circle, from heroes to local histories, tales of magic and metamorphosis, panegyrics and protests, fables of planting and felling, and on through forests of wood carvings, masts, musical instruments and furniture, until I am back in the same room, surrounded by familiar things. They are never quite the same

AGNES MILLER PARKER, ACORNS, BERRIES, CHESTNUTS, HAZELNUTS

afterwards: the table is no longer just a table. Oaks, like every species of tree, bristle with layers of meaning, forever undulating, opening, growing, fading, interleaving. The golden bough takes me to imagined futures, too, which spring just as abundantly from the arrested buddings at the end of each dried-out twig. Most of all, though, it compels me outside, even on the coldest, wettest days, to breathe in life from the nearest trees and generally take stock.

Well, perhaps not on the wettest days. The local clay, cracked and hard in August, becomes so waterlogged in winter that field gates are almost impassable and the sensation of cold mud seeping over the top of a boot can detract a little from the beauties of nature. It is only after rain, though, that the trees become translucent, every extremity glistening with tiny crystal balls. A January morning can be the best time of all for seeing trees, because it is when the leaves have been stripped away entirely that the graceful symmetry of an alder or the wispy cascade of a silver birch is clearest. It is easier to spot things that are normally hidden away, too: last year's nest of tangled sticks blotting the line of an upper branch, or an outcrop of creamy fungus, like a stand of ghostly parasols, at the foot of a nettle-free trunk. Even when the day is drenched in grey and there is hardly anything to be seen, the ashes will still be brandishing their black buds as if pointing towards some higher region of colour and light.

In spring, you can feel life stirring in the barest twigs and the silhouetted catkins look as if a diminutive duck has run across the sky. One day the twigs are just beginning to thicken and brighten and bulge; by the next they are covered in pincer-paired leaves and pale, lime-white or pink-tinged blossom. There is nothing tentative about these vernal explosions. When the days are longer, it is all sap and fresh smells, and the liquid calls of birds hidden in the drifts of thicker foliage. The bark has been through it all before, but the craggy faces of ageing willows and the peeling skins of cherry trees seem less pinched in the bright light. By early November, when it is all dank and dark, the woods have a different taste, which does not quite match the ember-fall, sugar-brown shaken leaves.

I have always begun to feel suffocated after too much time indoors. The arboreal impulse is all outwards, into fresh air. Each tree is a mass of little bursts of energy, seemingly at odds with each other, though remarkably harmonious overall. Each variety has its own character and calendar, standing ready to join in the waves of spreading greens or golds at just the right time. 'Oak before ash, we're in for a splash;

ash before oak, we're in for a soak.' Whoever came up with this old rhyme was evidently more interested in cheering people up than in predicting the weather, because the ash seldom ventures into leaf before the oak.

It is not just the seasonal cycle of shifting colours that makes a wood so compelling: the very same trees can appear quite differently on consecutive days, or even at different times of the same day. For Samuel Taylor Coleridge, left behind to recover from a mishap with boiling milk while his friends went off for a country walk, the garden seat under the lime seemed like a prison – until he began to imagine what they were enjoying and, with that, his own lime-tree bower turned into a mass of 'broad and sunny' leaves, dazzling, dappling, liberating and uplifting.

Tree transformations are not merely a matter of the beholder's mood, of course. Claude Monet would set up three canvases side by side, moving from one to another with the light, in his attempts to capture nature's colours truthfully. His sequence of *Poplar* paintings shows a line of trees snaking along the River Epte in bright sunlight, in strong wind, or in the more subdued tones of an overcast day. These were trees whose fascination could not be withered by age, nor staled by custom, like the pine trees at Mont St Victoire which Paul Cezanne painted again and again, never tiring of their familiar and yet still oddly elusive forms.

All kinds of tree can reveal unexpected internal connections. The smell of cypress in the rain, or a drift of blossom on a warm spring day, can waft us back to moments on wet pavements or under that old, half-forgotten, pear tree: to those indelible marks of personal history that lie unrecorded by camera or anecdote. Any horse chestnut with a strong, spreading lower branch and puckered trunk can take me back to one I used to climb as a child and ride like a cantering horse or a boat skimming the waves. We moved house frequently, so I do not know whether the tree is still standing, but like many others before and since, it seeded itself in my mind and is there, ready to be shaken into imaginative leaf if prompted.

THE TREE OF KNOWLEDGE

I very rarely seek out trees for old times' sake, though. I like them for themselves. The commonest of trees, especially, possess the powerful appeal of things that just grow because they must – it is what trees do.

Their indifference to the moods of those whose lives come and go beneath their sway has helped to secure certain trees a special place in human society nevertheless. In some cultures, they have marked both the starting point and the centre – both the Tree of Life and the Tree of Knowledge were said to stand at the very heart of the Garden of Eden. In Maori culture, the God of the Forests, son of the sky and earth, is the huge, 2,000-year-old kauri tree, Tāne Mahuta, which still towers above the Waipoua Forest. In Viking mythology, the entire universe was understood in terms of a great ash tree, Yggdrasil, whose branches were home to the gods and whose multiple roots stretched outwards to Jötenheim, land of the Giants, and down to Niflheim, region of the dead. (This has always puzzled me since, inspired by Old Norse literature, I travelled to Iceland and found not a single tree in the entire country.) The Druids of ancient Europe gathered mistletoe for their sacred rites, performed in the natural temple of a great oak forest. In Greece, too, priests devoted to Zeus interpreted the oracular rustle of oak or beech leaves at the shrine of Dodona. The mistletoe sprigs still sold at Christmas markets and the wind chimes suspended from today's ornamental shrubs may well hark back to the sacred groves of our distant ancestors.

Siddhartha Gautama found enlightenment during his quiet meditation under the Bodhi tree and, ever since, his followers have gone on planting the same fig trees (*Ficus religiosa*) at Buddhist monasteries. I was given one of their heart-shaped leaves in Nepal, after it dropped from a stout, tubular tree on a steep hillside, dwarfed, but not diminished, by the Annapurna. I hope it survived the earthquake. Sacred trees have a tendency to recover, helped along by those for whom they mean so much.

Jesus rode on a donkey from the Mount of Olives along a palm-strewn road, to be arrested in a garden and crucified on a wooden cross. His parables abound with figurative fig trees, mustard seeds

and vineyards. Craftsmen in medieval Europe, inspired by the New Testament, carved intricate leaves into screens and misericords for churches with naves modelled on the smooth trunks and soaring branches of mature trees. When Antoni Gaudí designed the soaring modern basilica of Sagrada Família in Barcelona, he drew inspiration from the Bible, the architectural tradition of Europe and the abundant vegetation of Catalonia. Trees seem to speak a universal language of form, but they are rooted in local regions with their own soil, climate and associated species. My first real sense of this came in the shape of a small, green paperweight, brought home by my eldest sister from the other side of the world. It looked to me like a family of sea snails, but when I visited her a few years later I found that it was modelled on the unfurling frond of a silver tree fern – the New Zealand symbol of new life.

From the banyan tree in India to the African baobab, from the biblical Tree of Life to Charles Darwin's evolutionary diagram of universal existence, trees have offered numerous patterns of connection, survival and understanding. Unlike a flow chart, which implies movement in one direction, a tree offers multiple possibilities – up, down, forwards, backwards, hierarchical, carnivalesque. A family tree is a kind of instinctive metaphor for linking blood relations through proliferating generations, but individual members may find themselves pictured as branches, leaves or roots. Older representations of genealogical trees often show successive kings or chiefs arranged vertically along the stout trunk of an oak, surrounded by their leafy wives, daughters and younger sons. Nowadays, the 'trunk' is more likely to be the person engaged in the hard task of digging through online records, a sense of self sprouting fresh shoots with every newly uncovered birth or marriage certificate. My uncle's extensive family records trace some branches back through two centuries, but as they are all on my mother's side, the tree will look rather lopsided until someone sets to work on the paternal ancestors.

Families and nations grow like healthy, well-balanced trees – or so we like to imagine. As the tree flourishes, so do we, which is what makes these natural, home-grown phenomena such popular images of collective identity. Mature trees are frequently evoked as symbols of longevity against the odds, but they are remarkably adaptable to fresh human demands. New associations can be grafted on, which may gradually grow into the central meaning, while older ideas are shed altogether. With the fall of the Ottoman Empire, Lebanon, free to choose its own national flag once more, adopted the image of the evergreen cedar against a white background. Even under subsequent French dominion, the tree still occupied the middle of the Lebanese tricolor and has continued to stand for the modern republic, framed by the horizontal red bands of the independent state. When Canada became independent, there was widespread feeling that the cultural legacy of the British Empire should be subdued. New images were needed, and yet, the impulse to express independence was twinned with a desire for long-term stability. Trees, with their indigenous credentials and perpetual habits of renewal, were perfect for the purpose and so, after considerable debate, George Stanley's striking red and white design was formally adopted for the new flag, making the maple leaf the official symbol of Canada.

These national icons are always based on distinctive species. The cedar is Lebanon's most famous native, recognised across the world for its majestic stature, not to mention distinguished appearances in the Old Testament. For the Lebanese people, it stands for peace and eternity and serves as a symbol of enduring hope in a very small country with a long history of war and invasion. Canada is home to at least ten different kinds of native maple, so right across the vast land mass people can recognise their national tree in the woods as well as on the flags. Not all the varieties flush quite so brightly in the fall, but both the sugar maple and the red maple can turn wooded slopes into dazzling clouds of scarlet and crimson, equal to the most gorgeous sunset.

In Britain, long before the modern idea of nationhood emerged, people were naming their homes after the local flora. In Norfolk,

North and South Elmham are villages where elm trees once grew, while Salle was the place for willows (from the Old English word for willow, *salh*). Water-loving willows also inspired Willoughby in Lincolnshire, Wilden in Bedfordshire, Willey in Shropshire, Willitoft in Yorkshire and, rather less obviously, South Zeal in Devon (another descendant of *salh*). In the Lake District, where regular rainfall ensures a flourishing variety, we can find Yewdale, Birks Bridge, Derwent and Applethwaite. The origins of Hazelwood in Derbyshire, Mountain Ash in the Rhondda or Poplar in Greater London are not very difficult to work out, but neither Edgcott nor Boxted signals their ancient association with oaks and beeches quite so readily.

The trees in many of these places were felled long ago to make way for houses, but memories of a lost terrain often remain. On a modern estate not far from here, you turn from Hawthorn Way into Sycamore Crescent before ending up in Hornbeam Close – as if finding your way through the network of disconcertingly similar roads is not so very different from a woodland walk. I like to think that the street names have something to do with the natural forms – that the choice for the crescent was inspired by the symmetrical wings of a sycamore seed, the cul-de-sac after the balloon shape of a full-grown hornbeam. In the land of the newbuild, these street names may well stem from a deeply buried, Antaeus-like need to remain in touch with the earth.

People have always gathered around trees, but especially any that are out of the ordinary – taller, fatter, older, or somehow deformed. They are obvious natural landmarks – easily recognised and so essential to the ancient grammar of any place. The rural calendar was once marked by trees: old elms, where May dances took place and 'gospel oaks' where people would pause to offer prayers during the annual ceremony of beating the parish bounds. A 'clipping tree' was usually a mature broadleaf such as an elm, oak, chestnut or sycamore, with shade enough for the strenuous work of shearing sheep in early summer. Trees were integral to the community – familiar to everyone, at times, almost familial.

And they still are. The Kent County Cricket Ground is famous for the special rules demanded by its special feature: the St Lawrence Lime Tree. Until recently, the challenge for batsmen was to whack a ball right over the lime rather than just the line. When the old tree was eventually stricken with disease in 1999, a sapling was nurtured in readiness, but as the great lime split apart in high winds only five years later, the youthful substitute had to be placed just outside the boundary to avoid being flattened by a first-class hit. The allure of the lofty lime tree, with its light-catching leaves and tiny sunburst blossoms, lingers long. In Slovenia, the vast Najevnik linden tree, the legendary site where retreating Turkish invaders were said to have left their golden spoons, is still a venue for major events, including an annual political assembly. The ancient lime tree in Kaditz, near Dresden, too, already monumental when Goethe visited it, has survived fire and bombardment to become a local gathering point for carol singing, folk festivals and, in 2010, an outdoor screening of the World Cup.

Cultural landscapes are traditionally studded with significant trees, as local guides like to boast: 'This is the very oak tree where William Wallace rallied his men, . . . where Robin Hood outwitted the

THE ANCIENT LIME AT KADITZ IN 1837

Sheriff of Nottingham, . . . where Dick Turpin borrowed and (somewhat improbably) later returned a bag of gold . . .' Or, moving further afield, there is the Caesarboom, a vast Belgian yew tree under which Julius Caesar is said to have taken a brief rest from his conquest of Europe, or the striking outline of the Lone Pine, marking the devastation at Gallipoli, or the triumphant Freetown Cotton Tree, where the first freed slaves held services of thanksgiving in Sierra Leone.

Some trees retain their celebrity status not as witnesses to something, but as seedlings of someone. On the South Lawn at the White House there is a magnolia planted by President Andrew Jackson, and in Massachusetts an even more venerable pear tree, which was set by the first governor of the state, John Endicott. All over the Scottish Borders are mature trees planted by Walter Scott, who was almost as active with a spade as with a pen. These local features often testify to a less familiar side of a legendary figure, like the oak tree in County Clare, apparently planted by the High King and warrior chief of Ireland, Brian Boru. Planting is often a foundational act, and a public pledge of future prosperity.

Even trees that have long since disappeared can still survive in memory. The Logan Elm in Ohio, where Chief Logan made his stirring lament for his massacred tribe, is marked by a commemorative stone since the old tree succumbed to Dutch elm disease, before collapsing in a storm in the 1960s. The Hart Horn Tree near Penrith, with two skulls and a set of antlers nailed to the trunk, kept alive for centuries the memory of a stag chased across the Border into Scotland and back again by a single hound, until they both died of exhaustion in Whinfell Forest. Wordsworth was unimpressed by those responsible – 'High was the trophy hung with pitiless pride,' he declaimed – but he was still moved by the way in which an old tree could sustain local memories for so long.

Old, hollow trees are especially prone to being filled with tales of the bold characters who have hidden themselves inside and, as the years pass, the hollowed becomes hallowed. Or haunted. Headless horsemen and phantom hunts tend to gallop through the darkest forests, along with the ghosts of ravished maidens, murdered brides and lost children. The great forests of Germany, dense with ancient oak and evergreens, sprouted fairy stories as readily as fir cones and spruce needles, inspiring the Grimm brothers to gather them up to frighten children far and wide. Spectral stories can also thrive in lighter conditions. The mixed hazel, bird cherry and ash in Wayland Woods in Norfolk allow enough light for a sea of spring bluebells and yet still harbour memories of the Babes in the Wood and their melancholy fate. The oldest living tree in Manhattan is the Hangman's Elm in Washington Square, which grew close to the site

HANGMAN'S ELM IN WASHINGTON SQUARE, NEW YORK

of nineteenth-century executions. In Crieff, Perthshire, which takes its name from the Gaelic word for tree, *craoibh*, the tree in question was the gallows where those who had behaved especially badly at the notorious cattle markets were executed and left on display.

Trees can still mean trauma for those injured in chainsaw accidents or crushed by the sudden fall of a heavy branch, or for those whose lives are destroyed by forest fires or a high-speed collision with a roadside tree. Fatal accidents are often commemorated by a sapling planted close by: a young willow beside the crossroads, a hornbeam near a building site. The tree grows, not to replace the person whose life has been truncated, but to sustain the memory and create a place of quiet contemplation, offering some solace to survivors. These living memorials are also quiet, self-effacing expressions of faith in the future.

During the First and Second World Wars, soldiers frequently spent their spare time carving the names of fallen friends on the bark of trees. The grey surface of a beech trunk was more amenable to a pocketknife than stone and, as the tree grew, so the letters and dates expanded, softening as the years passed. As Chantel Summerfield's recent discovery of wartime 'arborglyphs' – or tree graffiti – on Salisbury Plain has revealed, soldiers' thoughts ran just as often on absent wives and girlfriends, whose initials and other distinguishing attributes were driven into the smooth bark by memory and desire.

Shakespeare had a great deal of fun with the smitten hero of *As You Like It*, who vents his frustrations by carving Rosalind's name on every tree in the Forest of Arden. Unlike Orlando, the poet Andrew Marvell claimed to prefer the trees:

> Fond lovers, cruel as their flame,
> Cut in these trees their mistress' name:
> Little, alas, they know or heed
> How far these beauties hers exceed!
> Fair trees, whereso'er your barks I wound,
> No name shall but your own be found.

Though reticent about the reasons for his inclination, Marvell was quick to invoke the Greek gods as inspiring models: 'Apollo hunted Daphne so / Only that she might laurel grow'. This is probably not quite in line with what the twentieth-century soldiers, whittling a pin-up into the bark before going into action, were thinking, but it is a memorable response to the ancient tales of violent pursuit that so often end in the transformation of an unfortunate young woman into a tree.

In the eighteenth century, people even started commissioning portraits of trees. When the 3rd Earl of Bute retired from public life to improve his estate at Luton Park, he invited Paul Sandby to come and paint his finest specimens. Sandby's striking portrait of an old ash pollard in Luton Park is one of the earliest depictions of a tree in its own right – neither as background, nor framing device, nor accessory for figures, but right there at the centre, commanding full attention with its strange mass of smooth trunks, springing from a single stool.

Traditional techniques of forest management, such as coppicing and pollarding, which involves cutting through the trunk to encourage the regrowth of multiple shoots, had the practical benefit of providing poles, posts and cattle feed. It also prolonged the life of the tree: some of the voluptuous pollards in Epping Forest are hundreds of years old. When a pollard was left to grow, the multi-trunked trees that sprang from the top of the trunk thickened into a distinctive beauty, as appealing to artists and writers as to builders, shipwrights or furniture makers.

Magnificent trees are natural wonders within the scope of everyone, irrespective of income or education. For the visionary, even the most ordinary trees can offer a sudden glimpse of the divine, as William Blake found in an early, life-changing experience at Peckham Rye, and Stanley Spencer would affirm again and again in the fields around his home at Cookham. The extraordinariness of ordinary trees is brilliantly captured by David Hockney in *Twenty-Five Big Trees between Bridlington School and Morrison's Supermarket on the*

Bessingby Road in the Semi-Egyptian Style, 2009, a huge frieze constructed from photographs of individual trees, which evokes the experience of walking along a tree-lined road. It is a powerful affirmation of the transformative presence of trees in modern life. Anyone can walk along the Bessingby Road; many people do on a daily basis. At every step, even the most solitary pedestrian is attended by magnificence – a grand parade of green, gold or grey figures, depending on the season. In any town, in any city, the eternal turn of the natural cycle is gently registered in tree-lined roads, which everyone shares though not everyone notices.

It is often only when local trees are on the verge of disappearance that people begin to realise just how much they mean. As Edward Thomas put it in 'First Known when Lost', a poem prompted by the sight of a woodman lopping the last of a small copse of local willows, 'I never had noticed it until / 'Twas gone'. The sense of loss prompted by tree-felling has been echoing through British culture for centuries. Plans for new building projects that are known to put green sites at risk provoke passionate protests. Whether the threat comes from new roads, High Speed rail, supermarkets or plant pathogens, the urge to defend the environment, to stand up for ancient rights and save the trees for future generations is widely felt. The opposition to the draft bill proposing to dispose of many of Britain's forests in 2012 was symptomatic of a long-standing and widely shared attachment to trees. There may be less reliance on wood-gathering for domestic fuel than was once the case, but the need to live in a land filled with trees remains as deep as ever. Forests, woods, individual trees all stir profound emotions, which may lie quietly dormant or rise rapidly into full consciousness.

The success of Thomas Pakenham's *Meetings with Remarkable Trees*, first published in 1996 and frequently reprinted, shows that enthusiasm for trees is flourishing. Among the celebrations for Queen Elizabeth's Golden Jubilee in 2002 was the identification of the fifty Great British Trees, all now marked by a suitably green commemorative plaque. The Ancient Tree Hunt, supported by the

British government's 'Keepers of Time' policy, with the aim of recording every significant tree in the British Isles, attracts numerous spotters. As screens make virtual access to the natural world ever easier, it seems that people become more and more determined to see, smell and touch the real thing.

Although wood and woodlands may have receded from modern, man-made, mass-produced urban economies, trees are still essential to life. Cities no longer rely on hornbeam logs for heating ovens, oak spars for the frames of houses or willow-wood for carriages, but everyone needs oxygen to survive. The respiratory systems of trees, which feed on carbon dioxide and exhale oxygen, complement those of humans so well that every tree is really a tree of life. Industrial logging in the rainforests triggers anxieties about tree loss on an international scale – but the looming prospect of global clearance is beginning to prompt agreements to act before it is too late. As awareness of the less welcome implications of a warming planet dawns, the modern world is turning once again to sustainable supplies. What better resource for humankind is there than its age-old companion, the tree?

This book is a personal exploration of the meaning of trees. Inspiration has come directly from the trees I have been lucky enough to see for myself, but its roots lie buried in earlier, often largely unconscious, encounters. As a child 'helping' my grandfather with his huge, smoking autumnal bonfires, I didn't pause to think about which tree each of the soggy, brown leaves had fallen from, any more than I worried about what kind of woodlands we walked through with the dog. I had no idea what kind of wood had gone into the making of a pendant my mother gave me; I just liked to run my fingers over the smooth, polished surface. All these experiences were probably acting as a kind of mental leaf-mould nevertheless, piling up silently inside.

Personal associations are often coloured by the accidental. My most vivid memory of the great grandparental bonfire includes a family of hedgehogs, discovered and ceremoniously deposited on the lawn by our bull terrier, minutes before the match was lit. To a small child, it was not clear whether the hedgehogs or the dog were sustaining more damage – and in the event, all were quite unscathed – but it made me very wary of piles of dead leaves for years afterwards. My first (and, for many years, only) experience of tree planting, as a teenage volunteer on an eco-agricultural project, left a very damp legacy – not because of the mud and puddles that seemed likely to drown the forlorn saplings, but because the drive home involved abandoning the car in the middle of a ford as winter water rushed in through the doors.

This is one reason why books are so helpful – stories and poems crystallise other people's experience and so help to subdue and rectify, as well as enhance, your own. The heightened responses of poets, prose writers and painters have helped me to see things in a new, often brighter, light, so much of what follows has grown from the leaves of books as well as trees. I will not begin to catalogue debts to favourite writers here, but many have helped to define the different kinds of tree that feature in each of the chapters below.

The meanings of trees are ringed with myths and histories, growing thicker as the years go by, but the heartwood remains hard and real. My understanding of the physical facts about trees has been helped immeasurably by many of the green giants of modern woodlands: Oliver Rackham, Richard Mabey, Roger Deakin, R. H. Richens and Gabriel Hemery (not to mention early authorities such as John Evelyn, William Gilpin, John Loudon or Walter Jackson Bean). This book has developed from the three series of 'The Meaning of Trees', which I wrote and delivered for Radio 3, *The Essay*. Special thanks are due, accordingly, to Turan Ali and Emma Horrell of Bona Broadcasting, to the Radio 3 editor, Matthew Dodd, and to the BBC. Preparation for this book has involved numerous field trips to significant trees, so I would like to thank my family not only for their patience in tolerating frequent expeditions and detours, but also for their support for the project and gifts of various woodland books. Robin Robbins first alerted me to the work of the Woodland Trust, which has been a great source of inspiration. I am also indebted to a number of people for various kinds of help during the course of writing the book, including Ann Blanchard, Ben Brice, John Cook, Jeff Cowton, Peter Dale, Jessica Fay, Linda Hart, Daniel Kurowski, Karen Mason, Andrew McNeillie, Kevin De Ornellas, Franklyn Prochaska, Jos Smith, Gill Stafford and Chantel Summerfield. I would like to record my gratitude to each of them. For enthusiasm, expertise, good sense and guidance, I would like to thank Heather McCallum, Melissa Bond, Rachael Lonsdale, Steve Kent and all at Yale University Press.

If *The Long, Long Life of Trees* springs from a sense of wonder at the physical beauty of these natural phenomena, their survival over centuries and the cultural associations that have grown up among them, it also looks forwards to a time in which the saplings being planted today will turn into the great trees of future generations. If anyone reading this book is moved to put it down and go in search of a tree or a spade, it will have done its work.

YEW

※

I<small>F</small> someone says 'tree', this is unlikely to be the first kind that springs to mind. It is not that yew trees are unfamiliar, but they are nothing like the simple idea of a round, green blob on a brown stalk that most of us internalise early in life, when learning what's what in the world. Yews are dark from crown to ground and come in all sorts of shapes and sizes. I have seen yew trees swelling up from the earth like great green sea anemones, their upper fronds bursting into the sky, and others whose branches hang down limply, more like battered old umbrellas. Yews can resemble a collection of tightly rolled umbrellas, too, or a group of green church spires standing far too close together. These are the Irish yews (*Taxus baccata* 'Fastigiata'), always more inclined to look up than down. The more common European yews (*Taxus baccata*) are generally more diverse in form. Some have soft, smooth limbs that intertwine companionably; some, all hairy and independent, keep their rough needles close to their chests. Yew trees can be so dense that not a blink of daylight filters through, while others open out beneath their canopies to reveal loose bundles of pipe-like trunks. Young, slim yews hold out their branches elegantly, as if getting ready to dance, but their older relations sometimes have such spreading waistlines that they seem in danger of complete internal collapse. Yews will band together in circles, keeping their soft rusty carpets hidden, but they also stand in splendid isolation, high on a limestone cliff or darkening the corner of a quiet field. So familiar, and yet so various – is this why yew trees have always seemed so troubling?

YEW TOPIARY AT LEVENS HALL

The Romans kept them firmly in line, in rows of neatly trimmed green obelisks or angular animals standing to attention along straight streets. Renaissance Europe followed their lead, making these trees into dense mazes or planting them in geometric patterns for parterres. Yews provided a living building material, which meant walls could be grown in gardens and adorned by striking outdoor sculptures that the rain would refresh rather than erode. At Levens Hall in Cumbria, the topiary gardens, originally laid out in the 1690s, have gradually grown into a looking-glass world of fantastic forms: giant top hats and helter-skelters, startled mushrooms and stacking rings, birds and beehives, pyramids and chess pieces, an evergreen tea party of cups, cones, dark doughnuts and irregular jellies. It is a green dreamworld where everything is carefully clipped and controlled; an illusion of imaginative freedom, dependent on a great deal of hard work. Yews naturally assume very strange shapes, but here they can be found forming sturdy arches with solid emerald crowns on top. And yet, if you step inside the old yew hedge, the tangled wires of branch and root reveal the real source of energy beneath the spectacular scene.

During the eighteenth century, tastes turned to relish the wild and untameable and so the great estates were transformed to resemble more natural landscapes. Sculpted symmetrical gardens fell out of fashion, elaborate topiary was mostly grubbed up and the old yew walks cleared in favour of broad, spreading prospects. There were some survivors, like the thick, evergreen wall at Rousham, through which guests could slip from spacious vistas to secluded flower gardens, or the cliff-like hedge at Cirencester Park or the more mountainous cascades of yew at Powys Castle. The huge, distorted forms of the great 'Elephant Hedge' at Rockingham Castle, all the more disturbing for being so outmoded, would inspire Charles Dickens to create the Ghost Walk at Sir Leicester Dedlock's chronically unprogressive country house at Chesney Wold.

Of all trees, it is the yew that has been most apt to provoke unease, fear and even dread. In *The White Goddess*, Robert Graves pronounced it 'the death-tree in all European countries' and it is easy to see why. The toxicity of the yew is widely known – every part of the tree is poisonous, except the little red arils, which thrushes and blackbirds eat safely, unwittingly helping the tree to propagate as they fly and let fly. The dark, soft, shiny fringes of foliage are deadly: one reason why these trees were planted within the walls of churchyards was to protect horses or cattle in neighbouring fields from grazing on the inviting, but toxic, evergreen boughs.

Shakespeare described the yew as 'double-fatal', because its branches are poisonous and the bows crafted from them, lethal weapons of war. The Battle of Agincourt became famous in British history as a victory against the odds: a massive French army defeated by a small, but invincible, force of English and Welsh archers. The relative size of the armies was magnified by hindsight, but key to the myth was the deadly power of the yew longbow. The wood of a mature yew is amazingly strong because of its habit of slow growth, and yet yew wood is also flexible. For bowyers in the Middle Ages the

ENGLISH ARCHERS AT AGINCOURT

best part was where the dense, dark heartwood met the lighter, supple outer layer of sapwood, providing just the strength and springiness needed for their craft. The hard rain of hundreds of arrows, launched from these all-powerful bows, darkened the battlefields of medieval Europe – and must have been a terrifying sight for those who knew that their armour could not withstand the storm. A bodkin arrow was a sharp reminder of this deadly, needle-covered tree.

Widespread fears of the longbow also turned the archers into military targets: at Bannockburn, Robert the Bruce seized victory when the bowmen from south of the Border were hemmed in and fatally exposed, enabling the Scots to charge down on their warhorses to demolish the enemy's front line. Even off the battlefield, the life of the medieval archer was not especially enviable, because they lived in fear of the poison in their own longbows. If this no longer seems a very well-grounded anxiety, it does reveal the pervasive unease surrounding yew trees. Although the timber has long been prized and polished to reveal striking, red-gold, wood-waves for the finest cabinets, suspicions still linger. Some craftsmen are wary of turning yew into goblets, however enticing the rich, golden wood may look, for fear of toxic residue seeping into drinks.

The botanical name for the yew sounds rather alarming in itself: for English speakers, *Taxus* will conjure up thoughts of tax, taxing and toxicity. The *Oxford English Dictionary* defines 'taxin' as a 'resinous substance obtained from the leaves of a yew tree' and gives the first recorded use as 21 December 1907, in a newspaper report on a death that was unexplained until the post-mortem revealed a large quantity of yew needles in the stomach of the deceased. Exactly a century later, in December 2007, the Gardaí were similarly baffled by a Dublin suicide, until the forensic analysis revealed traces of the yew toxin, taxine B, in the victim's tea. The depressed and desperate are drawn to this tree. When Sylvia Plath was staying at Court Green, in one of the lowest periods of her short life, the yew in the nearby churchyard seemed the very image of the blackness in her mind. Her haunting poem 'The Moon and the Yew Tree' concludes, 'The message of the yew tree is blackness – blackness and silence'.

Dante placed suicides in a dark forest. He was no doubt thinking metaphorically, but the resonant image had physical foundations. He must have seen the ancient yew that still stands at Fonte Avellana when he stayed at the monastery, and travelled through the vast yew forests of medieval Italy. In the *Inferno*, a branch breaks from one of the dark, bushy trees, causing a torrent of words and dark blood. The only European tree that appears to bleed is the yew, which is capable of oozing a deep, red, remarkably bloodlike, sap. This phenomenon, still puzzling to plant scientists, has been drawing visitors for many years to the 'Bleeding Yew' in the churchyard of St Brynach's in the small Pembrokeshire village of Nevern. The weeping, crimson gash in this old tree has inspired stories of heavenly visions, ancient injustices, patriotic loyalties and universal peace.

The poisonous nature of the yew's foliage may be the most obvious reason for its associations with death, but many other plants are just as toxic and yet escape such a gloomy reputation. People happily give garden room to laburnum, foxgloves and lily of the valley with nothing but a little common sense as a guard. So what is it about the yew that triggers deeper fears? It may have more to do with the tree's appearance and location than with its natural toxicity. The dark silhouette in the graveyard has long been part of Western culture. Yews will grow in deep shade, thriving under the shadow cast by an old church, while their knotted, twisted trunks easily suggest contorted human figures. When painted or filmed imaginatively, a yew is guaranteed to create gloom. It is there in ghost stories and Gothic horror, in the melancholy graveside scene of a costume drama and the tensest moments of a crime series.

The yew's solemn presence looms through the gathering darkness of Gray's famous 'Elegy Written in a Country Churchyard' and casts its shade over Hardy's 'Lament' for his wife, Emma, sleeping for ever in her 'yew-arched bed'. When confronted by the devastating death of his closest friend, Arthur Hallam, Tennyson turned instinctively to the 'Old yew' in the churchyard to express his grief. In the shock of sudden bereavement, the tree seemed a monster, grasping at the tombstones, while its fibrous roots cocooned the body buried below. The young poet resented the 'old yew' for still being alive and active while the brilliant, beloved Hallam had been cut off years before his prime.

Tennyson was grappling with the peculiar grief of premature death in a society deeply unsettled by the dawning realisation that the earth was very much older than previously assumed and human beings very recent arrivals in its long history. The discovery of huge dinosaur fossils was making it hard to doubt that a world once understood primarily in relation to human history had, in reality, existed for millions of years, inhabited by creatures quite unknown to humankind. Tennyson's personal sense of loss was intensified by the new awareness of the extraordinary disparity between the lifespan of man and his habitat. Why should he be allotted such a brief moment, when a tree could live for a thousand years? As he worked through his profound grief, Tennyson came to see that far from being a figure of unremitting gloom, the yew, too, enjoyed a 'golden hour'. His depression gradually lifted with the returning spring, when even the dark old yew starts to splash itself in golden powder and join in the season of fresh life.

The longevity of the yew is not, after all, a sign of the injustices of nature and has often been celebrated as a source of wonder and symbol of hope. In Austria, yews were planted in village squares to bring good luck, while in Germany yew branches decorated homes at Yuletide long before spruce became the established Christmas tree. For the ancient Celts, yew trees were sacred; for the Romans and the Saxon peoples, they guaranteed safe passage into the next world. In Britain, yew trees continued to be planted in churchyards for practical reasons, but many were already in place long before the arrival of Christianity. The ancient yew at Watermillock in Cumbria probably predates the 'Old Church' of St Peter's, just as the tree at Much Marcle in Herefordshire, at the grand old age of 1,500, is several centuries older than St Bartholomew's Church where it stands. The first churches were often built next to yew trees rather than vice versa.

At Fountains Abbey in North Yorkshire, there is a group of ancient yews that were already big enough to give shelter to the monks while the abbey was being built in the twelfth century. This may be another symptom of the threatening nature of these trees, which, as sacred objects for pre-Christian society, were being carefully incorporated by the new religion; but it seems more likely that the old trees were regarded with reverence and continued to survive through successive centuries as symbols of the harmony between the Yorkshire monks and their yew trees. For these men, God was

YEW AT MUCH MARCLE

providing shelter and a natural house of prayer – and hence, no doubt, the safety of the yew trees at the abbey, even at periods when military demand for yew timber was at its height. For those whose lives were focused not on the present but on the life eternal, yew trees were reassuring companions. These trees furnished greenery for shrouds and funeral processions, because the unchanging evergreen, with its blood-red berries and shining leaves, immune to winter's depredations, was a symbol of everlasting life.

Yews are among the oldest living things in Europe. Given the veneration afforded to ancient cathedrals, castles and Roman remains, it is odd that the living survivors of antiquity are much less well known. The Ankerwyke Yew on the banks of the Thames at Runnymede was already a portly veteran when the Magna Carta was signed there in 1215. Three centuries later, the old tree stood as a silent witness to Henry VIII's courtship of Anne Boleyn. Some yews, like the tall tree in the courtyard of Skipton Castle or the splendid pair flanking the door of the parish church at Stow-on-the-Wold, are part of the very fabric of the building. The Great Yews of Wales Trail includes the Bleeding Yew at Nevern, the arching yew tunnel at Aberglasney and the Pulpit Yew at Nantglyn, where John Wesley is supposed to have climbed the steps up its great trunk in order to preach.

Great trees such as this were not objects of worship, but rather natural centres where people who followed creeds other than that of the established Church were free to congregate. On a visit to Cockermouth in 1653, George Fox met fellow Quakers at the huge Lorton yew. When Thomas Pakenham headed for the Lake District in search of this legendary tree, eulogised by Wordsworth as the 'Pride of Lorton Vale', he was struck by how little attention the massive yew commanded when he finally came across it in a field behind an old brewery. It had evidently long since vanished from the tourist radar screen, for as early as 1902, *The Methuen Guide to the Lakes* pronounced it a thing of the past. Even the most enormous trees are capable of disappearing from view – or at least from public consciousness, it seems. Rumours of its death probably began in the 1840s, when half of the great giant

THE ANKERWYKE YEW FROM JACOB STRUTT'S *SYLVA BRITANNICA*

succumbed to a ferocious storm and the remainder was nearly felled for timber.

If you visit Lorton, you can still find the yew beside Hope Beck, outlined against the distant hills and now identified by its own laminated board, diminishing the sense of achievement somewhat, though probably helping to secure the tree's future. The tree is a little lopsided still, but magnificent even in its semi-detached status. Two hundred years ago, the famous Lorton yew was described by Dorothy Wordsworth as the largest tree she had ever seen: 'We have many large ones in this country, but I have never yet seen one that would not be but as a branch to this.' This 'Patriarch of Yew trees' once seemed almost biblical in age and stature.

Some of the living yews in Britain are older than Stonehenge, older than the Pyramids – a thought that takes a moment or two to grasp. What has seemed a normal, unremarkable part of the landscape, common in both town and country, suddenly starts to magnify as you contemplate what has already happened beneath its shade, or its own potential to outlive all the surrounding buildings, roads and villages. Trees that were seedlings two or three thousand years ago were already vast by the time the Romans arrived, and this is why legend has it that Pontius Pilate, who was supposedly the child of a visiting envoy from Augustus Caesar, played beneath the great yew tree at Fortingall in Perthshire. The age of the great tree is the most persuasive part of this story, but the detail, implausible as it might be, does help to create a relative scale for these extraordinary natural phenomena. Wales, not to be outdone, has its own contenders for the oldest living inhabitants of the United Kingdom – the Defynnog yew in the Brecon Beacons and the Llangernyw yew, further north at Conwy, both of which may have been alive some five thousand years. There is no suggestion that Pontius Pilate was ever in the Brecon Beacons, but at the time when Caractacus was putting up resistance to the Roman occupation, the Defynnog yew would already have been sending down new roots in its regenerative cycle.

No one yet knows the precise age of the oldest trees, so estimates range disconcertingly over hundreds and even thousands of years. The Llangernyw yew may well have reached the grand age of five thousand, but there are counter arguments to suggest that it is a mere stripling of only 1,500 years. Unlike most of their relatively short-lived neighbours, yews have the rare capacity to regenerate, which makes determining age through rings of bark peculiarly problematic. Extremely slow growth means that the annual rings are often only a fraction of a millimetre apart – even a magnified section of wood looks more like the edge of a tightly bound book than an orderly, measurable sequence of brown markers. Such difficulties only add to the inscrutability of ancient yew trees.

As yews age, they begin to hollow, so some of the oldest trees are living shells, almost like wooden henges. This is another secret to their survival, since an empty, perforated tube is much less likely to be blown down in a gale than a solid, wooden column. Without a full complement of internal growth rings, however, dating through dendrochronology is impossible, and the yew's irregular growth patterns mean that even a partial section of the trunk or a branch will not deliver an accurate age for the whole tree. The yew's ancient association with everlasting life is more than sustained by its natural habits, for its extraordinary methods of regeneration mean that although some trees shed their soft, flaky bark after a few centuries, they can still continue to thrive because of the new roots that shoot down internally from crown to ground, before developing bark of their own. The inner bundles of tubular roots are hard to distinguish from the corrugated trunk of the original, making the delicate question of age more and more difficult.

An ancient yew, such as the enormous tree at Tandridge in Sussex, may look as if the great god Pan has hurled down his pipes in fury, leaving them to fall in different directions. The Ankerwyke Yew, on the other hand, is more akin to a gnarled rock face, opening slowly to reveal a cave of peat-brown stalactites and petrified creepers.

Even when an ancient yew finally yields to forces greater than itself, it is still reluctant to quit: in the village of Selborne in Hampshire, the tree in the churchyard that Gilbert White observed on a daily basis was blown over in the gales of January 1990. Though little more than a stump remains, a sapling from the ancient tree has been planted nearby, while the great remnant of Selborne's oldest resident now stands there like a world in itself – with its bare wooden cliffs and strange ravines of bushy leaves and greenery.

To find the age of yews, we need written records as much as scientific analysis. Naturalists have been measuring the girth of yews for centuries, and so we can now compare early records with the size of the trees today. The Fortingall Yew in Glen Lyon had swollen to 'fifty-six feet and a half' when the Welsh antiquarian and naturalist Thomas Pennant made his Scottish tour in 1769. It is just as broad today, though so fragmented as to seem almost like a group of smaller trees. Growth charts for yews certainly put NHS norms for human development into perspective – the arborealist Allen Meredith has estimated that a circumference of over ten metres means a yew of at least two and a half thousand years. Probably. Paradoxically, while longevity is very likely the yew's most remarkable feature, these elusive trees are still impossible to date.

Old drawings, paintings, poems and prose descriptions can offer important insights into the life stories of ancient trees. Wordsworth's poem 'Yew Trees' opens with the Lorton yew but quickly moves over the Honister Pass to contemplate the ancient yews of Borrowdale. His description of their

> Huge trunks! – and each particular trunk a growth
> Of intertwisted fibres serpentine
> Upcoiling, and inveterately convolved

is a highly accurate account of these self-regenerating ancient trees, in language appropriately grand and archaic. Wordsworth is equally observant of the 'grassless floor of red-brown hue', which is perennially tinged by 'sheddings from the pining umbrage', though his use of the word 'pining' embraces not only the local dialect word for the process by which vegetation dries out, but also the more widely understood meaning of yearning or languishing. As a poet, he was responding to the distinctive physical features of the trees as well as to their cumulative, cultural meanings.

Wordsworth had travelled through Borrowdale with his brother, John, in 1799 and 1800, but by the time he published the poem in 1815, John had been dead for ten years. He was a captain in the merchant navy and went down with his ship just off the Dorset coast at Weymouth in 1805. For Wordsworth to be recalling the Borrowdale yews as the 'Fraternal Four' had a poignant personal dimension, since he himself was now one of only three brothers. This may explain why he saw in the 'sable roof' of the yew trees only 'unrejoicing berries', and imagined 'Death the Skeleton' and 'Time the Shadow' meeting in this natural temple.

The 'Fraternal Four' no longer stand as square as they did when Wordsworth visited them, because one was blown down in a gale in 1883. Their location continues to be marked on Ordnance Survey maps of the Northern Lakes, even though most of the walkers heading towards Scafell or Great Gable probably pass by without registering quite what lies hidden on the slope above the Derwent. The ancient yews are still the most enigmatic presences in this under-populated valley, silent as stone and yet exuding an air that is not uplifting, but neither quite melancholy, a stillness so deep that even breathing seems intrusive. When I tried to photograph the Borrowdale yews under the bright sunlight of an August afternoon, my camera broke.

Photographs often do preserve the personal histories of yews better than anything and some trees now survive only as images on old postcards. Antique photographs can show an ancient tree prior to the loss of a major branch or, indeed, after a significant modification. Victorian pictures of the Crowhurst yew in Surrey reveal that what is perhaps its most startling feature – the little door in the trunk – was already in place, though the tree was listing less violently than it does today. When I visited, the door was hanging open, as if the last tenant had gone out in a hurry but, once closed, the dignity of the ancient tree was rapidly regained. Above the door, where two branches must have fallen off at some point, the hollows are unsettlingly like great eye sockets, blind to the agitations of the present moment and yet, in that unfocused gaze, capable of seeing much further. The contemporary artist Tacita Dean has captured the strangeness of the Crowhurst yew in her striking portrait of the tree, adapted from an old postcard, but stripped of all context to create a quintessentially yew-like sense of timelessness.

Old trees often register the changing compulsions of their successive human companions nevertheless. When the yew at Crowhurst was being domesticated in the early nineteenth century, its craggy interior fitted out with neat table and chairs and the little door fixed to the trunk, a cannonball was uncovered. It had been lodged there during the Civil War and had lain undisturbed ever since. Yew trees are living monuments, shaped by human history and packed with all manner of surprising revelations.

The medicinal discovery of Taxol shows just how slow the yew has been to yield some of its secrets. In the 1960s, American scientists discovered a compound in yew trees that might be effective against certain kinds of cancer and, after extensive trials, Taxol was approved for chemotherapy in 1992. Far from being a death tree, the yew suddenly began to be rebranded as a tree of life. The Pacific Yew Act was passed in the United States to ensure responsible management of the yews on the West Coast, which until then had generally been regarded by lumber merchants as nothing more than 'trash trees'. Since the early 1990s, further research has led to new yew-based drugs for treating ovarian, breast and prostate cancer and the tree's therapeutic properties are still being explored.

The development of new anti-cancer treatments is a wonderful medical breakthrough, but the sudden demand from the pharmaceutical industry has had its darker aspect. Stripping the bark of

THE CROWHURST YEW

ancient yews provides raw materials for effective drugs, but it also kills the trees. The Himalayan yew, which has been growing in parts of Nepal, Afghanistan and India for centuries, is now an endangered species. Since taxanes can be extracted from the needles of yews, careful cultivation and harvesting, rather than the quick-fix solution of the chainsaw, would ensure long-term sustainable medical resources; but it would also mean smaller instant profits, which for countries in economic crisis are life-saving in a different way. The need for yew-based treatments is great, and eminently justifiable, so the impulse to fell has a humane as well as an economic dimension, making policy on the matter very complicated. But it does bring to mind a very long history of short-term thinking relating to the yew.

In medieval Europe, the demand for longbows led to the destruction of European yew forests, in an early version of the arms trade – with all its ironies. Yew wood imported from French forests might well return home to launch deadly arrows at the very people who had felled it. The yew trade was a major part of the European economy, but resources diminished rapidly and once the trees were gone, so too was the best material for weaponry and hunting. This is probably the reason why there are no ancient yews left in France and why the victorious bands of bowmen began to recede from history. A plantation of yew saplings would not be capable of equipping an army for many years – trees established under Richard III might only be ready to furnish bows for George III. By then the steel furnaces of the Midlands were providing materials for weapons capable of much more wholesale destruction.

Perhaps what is really troubling about the yew is not its dark façade, its multiple shapes or its poisonous needles, but rather its extreme longevity. Perhaps humankind finds it hard to cope with something that has seen so many regimes and policies come and go, and will carry on living into a future when all our aspirations have been forgotten. If all that is left of the people who planted some of our more venerable trees are their broken bowls and beakers, what will remain in two thousand years of the couples in today's garden centres, trundling their potted yews to the checkout, ready to plant a small hedge over the weekend? This is likely to be their longest-lasting legacy, but the thought of being less durable than a hedge may prove a little too humiliating.

The yew need not stand as a gloomy reminder of the fleeting quality of human existence: really it is a means of liberating us from limited perspectives. Something of ours can survive the centuries, just as the ancient yews of Fortingall, Llangernyw, Crowhurst or Ankerwyke have done. We do not know what else the yew may have hidden away inside, but one day we might. So what does the yew tree mean to humanity? I think it is much too early to say.

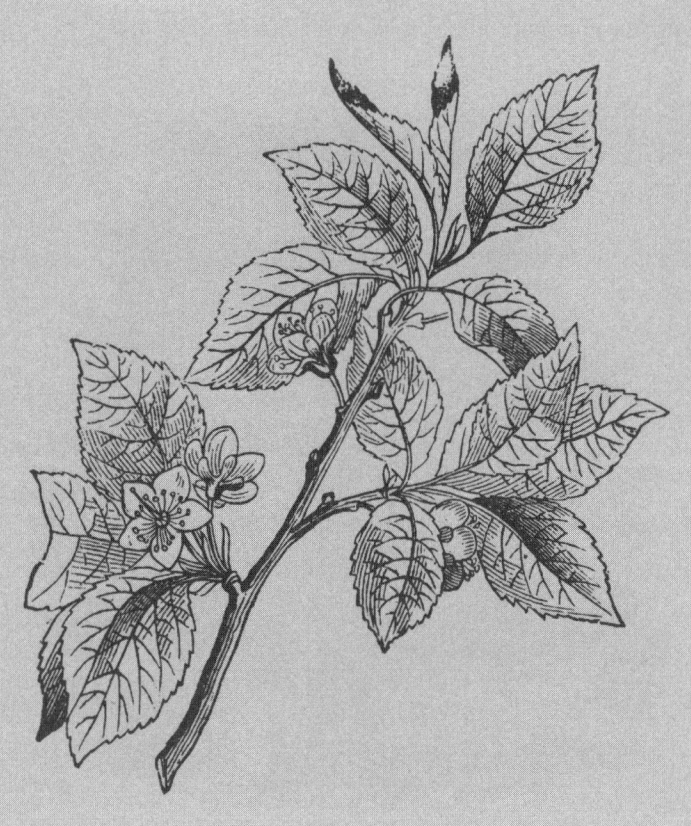

CHERRY

THE Tree Cathedral at Whipsnade was planted in the aftermath of the First World War. Captain Edmund Blyth, who survived the carnage, felt compelled, like many fellow survivors, to create a memorial to his lost companions, but at first could think of nothing that would begin to measure up to loss on such a scale. Some years after the Armistice, Blyth and his wife were visiting Liverpool and went to see the new cathedral, which had been under construction since 1904 and, though now consecrated, was still only half built. This was a work in progress, at once a lifetime project and an act of faith for the architect, Giles Gilbert Scott, whose vision was very slowly turning into a monumental statement for the people of Liverpool. On their way home through the Cotswolds, the Blyths' journey was halted by the sight of sunlight flashing suddenly onto an otherwise ordinary group of trees: at once a natural transfiguration and a moment of startling vision. Blyth realised that he too could create a great church, but instead of bricks and glass, it would be made of trees and sky. His open-air cathedral, more beautiful than any building, would never be complete because its growing columns would go on sending out arching branches filled with budding tracery. It would be a memorial to his particular friends and to the whole generation of young men cut down too soon, but it would also be a living articulation of shared faith in the future, planted in the spirit of hope and reconciliation. In 1927, Blyth had bought a farm in Whipsnade on the Dunstable Downs; now he knew what to do with the land.

THE EASTER CHAPEL AT WHIPSNADE

Eight decades later, Captain Blyth's saplings have reached the heights he hoped they would attain. Beside the tall trunks of the nave and the shimmering silver birch chancel stands the earliest of the chapels to be created by Blyth: the Easter Chapel, formed from cherry trees. Every year, this quiet space of contemplation, punctuated only by bare trunks and hanging branches, is suddenly filled with dazzling clouds, white as heaven against the pale spring sky. The annual transformation is most impressive in years when Easter falls late and the chapel lights up in celebration, but even when the festival occurs in March, the bare cherry trees still stand, unobtrusive but unmistakable, in patient anticipation of their moment of glory.

Wild cherry trees, too, light up woodland paths like heralds of the returning sun – a sudden suffusion of white, and then the flowers are gone again. A. E. Housman was being uncontroversial when he dubbed this tree 'the loveliest'. Despite some serious competition, the sight of a cherry in blossom is so breathtaking that, for a few days at least, no other tree can match it. Ted Hughes saw the arrival of cherry blossom as a kind of vernal party invitation, though it is ultimately a little disappointing because by the time the guests arrive, 'she rushes out past us, weeping, tattered and dirty'. These pretty flowers are so often wrecked by wind and rain, those unfailing attendants of the season, almost before anyone has been able to admire them. If the bright bloom of the wild cherry – or mazzard or gean, as it is also known – seems the very essence of an English spring, however, it is probably time to readjust our settings. Fresh, fleeting, fugitive, the cherry is one of the *world's* most popular – and most transient – trees.

The last week in March sees mounting excitement in Washington, DC, because of the multitude of flowering cherries around the Tidal Basin. As soon as the buds begin to stir, the cameras are there. Cherries are nature's celebrities and no one wants to miss out. It only takes three weeks for the flowers to appear, first as a gentle dusting of snow, then quickly rising into a glorious mist of pale pink, before falling in a shower of a thousand petals. Far from being a quiet communion with nature, blossom-watching American style is intense, sociable, and even surprisingly athletic, because the Annual Cherry Watch Ten-Miler is organised to coincide with the spectacular show. People gather to relish the cherry trees and cheer on the crowd of runners as they go panting past.

Cherry trees seem to move en masse, too – it is almost as if the word goes round and none wishes to be left behind. In Japan, cherry blossom is like the weather in Britain – an all-consuming, collective obsession at certain times of the year. The start of the Japanese spring is marked by blossom-viewing, or *hanami* festivals, with music, picnics and tea parties to accompany the stunning arboreal display. For a fortnight or so each year, the perfectly symmetrical form of Mount Fuji rises like an island from a sea of frothy white to be snapped by thousands of amateur photographers. The movement of the cherry blossom is like a sensational musical revue on tour, opening as the temperature rises on the southern island of

KATSUSHIKA HOKUSAI, *VIEW OF MOUNT FUJI*

Okinawa in January and making its way steadily northwards to the furthest extremities of the archipelago by May. Each orchard has its moment in the limelight for a few weeks every year.

The celebrated Washington cherry trees were, in fact, originally Japanese, arriving only in 1912, as a gift from the mayor of Tokyo. The First Lady of the United States, Helen Taft, and Viscountess Chinda, the wife of the Japanese ambassador, each planted a tree in ceremonial accord, leaving the other three thousand saplings for the city's groundsmen to sort out. (This was, in fact, a second attempt to establish Japanese sakura cherries in Washington – the first consignment had been infected with disease and had had to be destroyed.) Both of the original, matriarchal trees are still standing, along with masses of pretty descendants. After the Second World War, when the damage to relations between the United States and Japan had seemed beyond repair, a further diplomatic party of cherry trees eventually arrived to help with reconciliation and recovery.

Ornamental cherry trees, designated favourites of the emperor of Japan, are a defining feature of Japanese culture. Pink, white and perfect, the Japanese cherry tree has made its way across the world as a stylish figure in prints, on fabrics, porcelain and paper. The tree itself, when planted in congenial soil, has a special capacity to take root and flourish. The pretty little cherry tree accordingly spreads good will and harmony everywhere it goes – or almost everywhere. In post-war Korea, all the cherry trees that had been planted by the occupying Japanese force were torn up when the conflict ended and replaced by indigenous species. The trees stood for Japanese military power, an association strengthened by the cherry-blossom designs painted on Japanese bombers to symbolise the intensity and brevity of life. Some cherry trees have since been replanted, because South Korean botanists have now suggested that the ornamental cherry was originally native to their own country, but the issue, bound up with a bloody history of colonial relationships, remains contentious: China has recently registered a counter claim to be the original home of the cherry. The beauty of these trees does depend a little on the eye of the beholder.

Their special allure has also spurred on human intervention, making genealogies hard to trace, as the history of cross-breeding has resulted in numerous different cherries with subtly varied blossoms. Although the sato sakura cherry trees of Japan remained something of a national secret until the early twentieth century, inside the country horticulturalists had been developing exotic cultivars for so many years that it is now difficult to tell which are the native species and which the hybrids. The quintessentially Japanese yoshina, instantly recognisable from its profusion of pale, five-petal, golden-hearted flowers, for example, is almost certainly a nineteenth-century cross-breed. Judging by the latest experiments in the International Space Station, cherry trees are likely to keep on evolving rapidly: a tree grown from a seed sent into space some years ago is demonstrating astonishingly accelerated growth and producing buds four years earlier than normal. The cherry is clearly set to maintain its position as the fleetest of all.

The blooms of Japanese cherries are generally so much fuller than those of their European relations that they made quite a stir when first introduced into late Victorian Britain. Their slender twigs could whisk up egg-white flowers, lighter and airier than anything the garden-loving nation had seen before. Sakura trees were the landscape designer's answer to the sudden trend for parasols and kimonos and productions of *The Mikado*. Soon these brightly trunked, strikingly striped trees seemed to be marching up and down even the dullest street.

While the ornamental cherries arriving from the East were such an eye-catching novelty, the native cherries still possessed enduring attractions of their own. They do, famously, 'stand about the woodland ride, / Wearing white for Eastertide', but by July they have changed completely into green and red. It is the summer splendour of the home-grown cherry tree that has always secured a special place in British hearts, or, perhaps more accurately, in British mouths and stomachs.

Medieval castles and monasteries often cultivated cherry trees for their valuable supply of fruit. The orchards of the Middle Ages were once thought to be part of the rich legacy of the Roman occupation, but archaeologists working on a Bronze Age site in County Offaly have uncovered the remains of prehistoric cherries. The ancient people of Ireland were evidently enjoying cherry feasts long before the Romans arrived in England with their Mediterranean cuisine. The fruit of this tree is a culinary cornucopia. Sweet cherries are delicious straight from the tree, sour cherries just as tasty when cooked in pies and puddings. Cherries can be bottled in brandy or baked into cakes, or *clafouti*, crêpes or *Kirschtorte*. In many countries where the trees are familiar natives, cherries contribute to main courses, too, offsetting the fatty juices of roast duck or giving a fruity flavour to saffron rice. Kirsch and maraschino liqueurs can also capture the elusive essence of morello and marasca cherries, and keep their invigorating spirit to hand for years.

Bright red, sticky, glacé cherries may seem the epitome of an age where the supermarket reigns supreme, but cherries preserved in sugar were served to the Tudor monarchy. Henry VIII was so keen on these succulent little orbs that he ordered the royal fruiterer to plant huge orchards of cherry trees, turning Kent into the Garden of England. The county retained something of this Edenic character for centuries and older Kentish residents still remember the mass of magnificent trees and the enormous ladders that came out for the cherry harvest every summer. After the Second World War, however, Britain's cherry orchards went into catastrophic decline, not helped by the advance of the global giants, Turkey, the United States and Germany, who now dominate the international market. An astonishing 90 per cent of British cherry orchards disappeared in a couple of decades. By the 1970s, most people were much more likely to be driving a Datsun Cherry than planting cherry trees. In the modern urban world, where a ladder has become a dangerous source of potential compensation claims, the officially sanctioned cherry-picker now carries only an elegiac memory of English orchards and the natural rhythm of the seasons.

In response to this melancholy turn, there have been serious attempts to recover the cherry's fortunes in recent years, strengthened by the development of dwarf trees and polytunnels. These innovations may not enhance the cherry tree's traditional charms, but they are designed to maximise its assets. Steady demand from high-end furniture manufacturers for the dense, richly coloured wood of the cherry is also encouraging sustainable plantations. So valuable, in fact, is cherry wood that the location of the timber trees is often kept very quiet. Fortunately the sociable nature of cherry trees has not been utterly suppressed: in the old growing areas from Kent across to Worcestershire, traditional summer cherry fairs are being revived to whip up new enthusiasm for locally grown cherries.

Some enterprising growers have even launched schemes that allow people to rent a cherry tree, which means you can view your very own blossom in the spring and then enjoy bunches of glossy black fruit fresh from the branch in July. Is this another sign

TITIAN, *THE MADONNA OF THE CHERRIES*

of over-hectic, part-time lifestyles – as those of us too busy or too impatient to nurture a cherry tree are quick to borrow someone else's for a small fee? Or is it an inspired way of restoring broken connections, between growers and consumers, freeze-wrapped punnets and living trees, human beings and Mother Earth?

The multi-faceted therapeutic qualities of the cherry tree should not be underestimated. Although Henry VIII may not be the best advertisement for the health benefits of cherries, they were a traditional remedy for gout, fever and post-viral malaise. This is probably because they are bursting with vitamins and red anthocyanins, as well as being high in fibre. Current research is exploring their anti-oxidant and anti-inflammatory potential, as well as their possible assistance in tackling obesity. At one time, the stalks were used as an infusion for bronchitis, anaemia and diarrhoea, but more recently, extracts from the bark of the African cherry have been shown to be effective in combating prostate problems. Unfortunately, the excessive bark-stripping that followed this discovery means that the treatment is no longer available. Cherry stones, on the other hand, are still plentiful and are now sold commercially as a loose stuffing for pain-relieving pillows. Since cherries are natural producers of melatonin, eating a few before your head hits the pillow will also help ensure a good night's sleep.

Cherry trees have always been associated not just with a healthy body, but also with the soul. In Christian tradition, the cherry is the fruit of paradise and the heavenly reward for a virtuous life. The clear white blossom is an obvious enough symbol of purity, but it is the fruit that appears more often in Renaissance paintings of the Virgin Mary. Carracci's tender painting, *The Virgin and the Sleeping Child*, shows her holding a finger to her lips to hush a cherubic John the Baptist while her own baby sleeps. A small bunch of cherries lies on a nearby table to symbolise his eventual heavenly destiny. In Titian's more famous depiction of *The Madonna of the Cherries*, Mary is holding a relatively modest cherry bough, but in another version of the subject, from the school of Leonardo, the entire background is a mass of glossy green leaves and even glossier red fruit. In the old Scottish poem 'The Cherry and the Slae', the spiritual pilgrim is drawn on by the heavenly cherry and its promise of eternal life, even though the worldly sloe is so much easier to reach.

Given the sanctity of the cherry, it may seem rather odd that one of the best-known stories about George Washington involves the wanton destruction of one of these trees. As every child knows, young George took a hatchet to his father's favourite cherry tree, but when the crime was discovered, instead of concealing his guilt, the future president stood up and confessed, 'Father, I cannot tell a lie.' In this strange tale, the innocent arboreal victim is quickly forgotten in the face of courageous adherence to Truth. The other problem with this moral is, of course, that the story is almost certainly untrue, and is now generally attributed to Washington's early biographer, Parson Weems. The cherry tree is a

GEORGE WASHINGTON AND THE CHERRY TREE

crucial character nonetheless, because its innocence, beauty and wholesomeness make the boy's destructive act seem far more shocking than if he had simply chopped down an unspecified 'tree'. The detail conjures up an awful scene of desecration as the avalanche of petals falls over the garden, though most modern American depictions go for a cascade of red fruit rather than blossom.

The fruit is, after all, the most obvious distinguishing feature of this tree. The two red circles hanging from their inverted V-shaped stalk are instantly recognisable as shorthand for the cherry. This is a symbol that evokes the gambling hall and the seaside arcade, where seeing rows of cherries means hitting the jackpot – a reward of a rather different kind from the heavenly promise held out in religious paintings.

EIGHTEENTH-CENTURY CHERRY SELLER

And there is, of course, another side to the innocent cherry tree. Cherry-patterned dresses smack of funfairs and flirtation, cherry-red lips seem to say 'Come and play'. Songwriters have often responded to the promise of cherries, whose appeal is intensified by the brevity of the season. 'Cherry ripe, cherry ripe', the cry of old street traders encouraging all comers to 'come buy, come buy', reinforces the sense of urgency surrounding the fruit of this tree. D. H. Lawrence, never given to understatement, sends the hero of *Sons and Lovers* up into a cherry tree, 'hung thick with scarlet and crimson drops', just as his protracted courtship is reaching crisis point. As Paul Morel tears off handful after handful of the 'sleek, cool-fleshed fruit', the cherries touch his ears and his neck, 'their chill fingertips sending a flash down his blood'. The cherry is the tree for sacred and profane love, and ripe cherries, round and enticing, are a treat for the senses as well as the soul.

Even cherry stones have prompted thoughts of nuptial bliss – or at least of eligible husbands. People have been counting the stones to the chant of 'Tinker, Tailor, Soldier, Sailor' – or variations on the theme – for three hundred years at least. It struck A. A. Milne that children might find other possibilities a little more exciting:

> And what about a Cowboy,
> Policeman, Jailer,
> Engine-driver,
> Or Pirate Chief?
> What about a Postman – or a Keeper at the Zoo?
> What about the Circus Man who lets the people through?

Milne was not the only one to question the old list of occupations during the mid-twentieth century; the old rhyme was also resurfacing to offer young women a more up-to-date register of bachelors: 'Soldier Brave, Sailor True, Dashing Airman, Oxford Blue, Skilled Physician, Curate Pale, Learned Judge, Squire so hale.'

As RAF stations sprang up across Britain in response to the threat from the Luftwaffe, they were planted with cherry trees. Many of these matching trees still stand spick and span in military order, though their girths are no longer quite so trim and the horizontal lenticels stretch a little more widely. The trees present an annual parade of plumed helmets nevertheless, to match the smart, white kerbs and gates. Were they planted as a symbol of

what was at stake in the garden of England, inspiring the young men who flew off into the clouds? Was the cherry still understood to be the tree of heavenly rewards? Or was the fleeting beauty of the blossom a reminder of the terrible transience of life? The question of whether or not life is a bowl of cherries has been exercising us for a very long time.

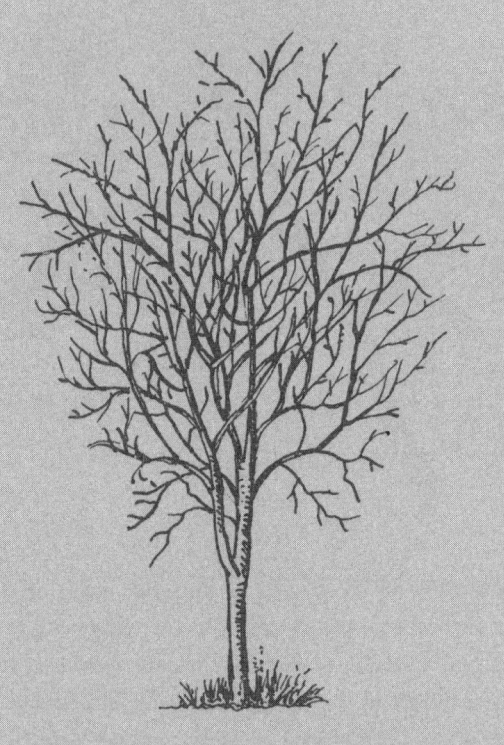

ROWAN

I T is one of those trees that garden experts recommend: easy to grow, good on all sorts of soil, low maintenance, and unlikely to get too large. This tree for all seasons earns its place in almost any garden, with a kaleidoscope of changing colours, turning from creamy spring blossom to pistachio summer green, before being showered in bright bunches of scarlet berries for the grand finale, an autumn spectacle of deep pink, coral and magenta. Bird lovers also approve of the rowan because it is such a favourite with blackbirds and thrushes, and so makes an excellent choice for those who like a lusty dawn chorus. No wonder it is so often described as 'useful' in practical handbooks and programmes offering advice on gardening. In the spring, you can even buy small rowan trees in the supermarket, complete with planting instructions and general advice: 'ideal for the smaller garden'. With so much to offer and so little demanded, it is not surprising that rowan trees are to be found in suburban streets and gardens all over Britain and Ireland. What is perhaps more puzzling is that the name usually given to these trees on the glossy cardboard labels is 'mountain ash'.

However familiar this neat, petite ornamental tree may be, its common name carries memories of a wilder ancestry. For this is a native of the northern hills and can still be found clinging to rocky faces in the Scottish Highlands, at altitudes of over two thousand feet. It is very often standing in isolation, silhouetted on a ridge against a clear winter sky, or giving focus to a vast brooding hillside, with its defiant clusters of berries or fiery autumn foliage. The tree

has been shaped by these bare mountain spots, its branches spreading out evenly from the top of the slim trunk, as if to help maintain balance on such a precarious perch. The rowan is as graceful as any of the waxwings or hawfinches that alight to gorge gratefully on its tart, red fruit. It is a tree of dual identity – at once a safe, respectable suburban accessory, which, providing those messy berries are promptly cleared away, should not upset the neighbours too much, but, at the same time, a free spirit, given to covering itself in glossy, scarlet beads and blushing deeply as it undresses. For Seamus Heaney, the rowan tree looked 'like a lip-sticked girl', while Iain Crichton Smith, recollecting his home in the Outer Hebrides, described the 'dewy rowans' wearing 'red dresses in these parishes of green'; though in the same poem, he also observed 'a weasel, sucking the throat of a hare beside a rowan tree'. Poets on either side of the Irish Sea acknowledge the startling, distracting, but also faintly disturbing, beauty of this wild, mysterious tree.

In the shared Celtic mythology of Ireland and the Scottish Highlands and Islands, the rowan is the tree of the gods, its berries a celestial delicacy. According to the old tales, when a berry accidentally fell to earth and grew into a tree that mortals might reach, the gods sent a one-eyed monster to stand guard, terrifying all comers. And yet, the tree was so compelling that when Princess Gráinne fell in love with Diarmuid and deserted Fionn Mac Cumhaill, Diarmuid had to slay the monster so that the lovers could take the magic rowan tree as their secret retreat. In another story, the great hero Cuchulain sees an omen of his approaching death when he comes across three witches roasting a dog on a rowan spit. This is a tree to treat with caution, for fear of what powers it might unleash.

The rowan goes by many names, each carrying an enticing hint, but none quite catching its full meaning. 'Rowan' recalls not the Celtic but the Viking influences in Scotland and the north – it derives from the Old Norse 'raudr' meaning red, for fairly obvious reasons. Pronunciation varies in different areas, and 'ruan' has given rise not only to 'rowan' and 'rowan tree', but also to 'roan tree', 'rauntry', 'round tree', 'rantry' and 'rowntree'. Since the nineteenth century, then, successions of sweet-toothed consumers have been just as delighted by Rowntrees as finches have been by the confectioner's etymological ancestor. The name of the 'Round Tree' suits the rosy, spherical fruit and also the smooth, often perfectly circular trunk, which seems to invite the clasp of human hands. In his poem 'Rowan', Andrew McNeillie describes the pang of nostalgia for the remote, mountainous regions of these islands provoked by placing his hands around the trunk of a tree planted in a suburban garden.

Though 'mountain ash' recalls a northern heritage, it now seems something of a misnomer because these trees are so common in the flatter south. Neither are they related to the ash tree. The confusion arose because of the similarity between the pinnate leaves – the rowan's sequence of feathery leaflets, opening from a central frond resemble those of the ash, although they are not as symmetrical.

The species are separate, nevertheless, the ash being a *Fraxinus* and the rowan a *Sorbus*. There is quite a history of mistaken identity, though, because the rowan used to be classified by baffled botanists with the *Pyrus* family, along with pear trees and apples. The rowan's current Latin alias is *Sorbus aucuparia*, which translates as the bird-catcher, because of those irresistibly juicy berries. It is also known popularly as the 'Fowler's service tree' in some areas and, in Germany, as *Vogelbeerbaum* for similar reasons.

In the south-west counties of England, the rowan is known as the quickbeam, quicken, quickenberry or quickenbeam, which has nothing to do with mountains or birds. Quickbeam has Germanic roots and reveals the old association in Anglo-Saxon England of rowan trees with life and with being alive, or 'quick' – a sense still familiar from 'the quick and the dead'. The Saxons used the quickbeam as a charm for unfruitful land, just as in German folk tradition the same tree of vitality was used for blessing cattle. Those blood-red berries and magenta boughs flush grey autumn days with assurances of life, despite the fading light and lengthening night.

The vigorous growth of the young trees meant that they were often cultivated as a shelter for the seedlings of later developers such as oaks, so this may have contributed to the rowan's reputation for both life and protectiveness. Then there are the berries, which are not only visually vibrant, but also valuable therapeutically, as a gargle for sore throats and tonsillitis, as prevention against scurvy, or a treatment for haemorrhoids. They are full of citric acid and natural sugars, as well as being strongly astringent. The rowan is both full of life and life-preserving, because the berries are gathered for culinary purposes, too – for making tarts, red jelly to eat with game, or for drying and grinding into flour.

When fermented, rowan berries can be made into cider, and in Wales they were used to produce a special beer, brewed according to a secret recipe – the original of which, as Mrs Grieve in her splendid *Modern Herbal* tells us, is now, unfortunately, lost. Much more

ROWAN LEAVES, FLOWER AND FRUIT

potent are the northern European spirits: rowan schnapps from Denmark and rowan-berry vodka from Poland. The trick with rowan berries is to pick them when they are ripe, just after the first frost has begun to mellow their sharp, acidic flavour, and then keep them frozen before submerging them in alcohol and leaving them in the dark. A shot of pale, red rowan schnapps will get the blood racing on even the longest, coldest days of a northern winter. The name of the quickbeam may, however, derive from quite different qualities, for the narrow fingers of its soft green leaves, never ceasing to quiver on their delicate branches, are constant reminders of life. The barely audible sound of the leaflets gave rise to yet another common name, 'the whispering tree', as apt for its suburban situations as for its wilder, hidden settings.

That there is more to the rowan than initially meets the eye is evident in the other names traditionally given to this tree – 'witchen', 'wicken', 'wiggen', 'witchwood'. While 'wicken' is probably a corruption of 'quicken', 'witchen' and 'witchwood' hint at magic and supernatural agency. In Celtic communities, the rowan was the wizard's tree. This does not mean that it was understood to be a sinister instrument of the powers of darkness: on the contrary, it was a form of protection against evil. In Scotland and Ireland, rowan trees were set near houses to guard the family inside from any threatening supernatural forces. At one time, rowans were being planted in almost every churchyard in Wales to help the dead on their passage to the next world and prevent spirits from being left behind to haunt the living.

When William Gilpin was studying the locations of British trees in the late eighteenth century, he noticed that there were often rowans growing near stone circles. While this probably fostered belief in the tree's ancient associations with Druid culture, it is likely that they were planted there far more recently. Rowans do not live for thousands of years, so those spotted at prehistoric sites must have been relative newcomers. They might have continued to seed naturally over the centuries, but it is also possible that they were planted by people anxious about any ancient supernatural powers still lurking at Avebury, Great Rollright or Castlerigg.

Belief in the protective power of the rowan was once very widespread, as evident in the old Scottish rhyme 'Rown-tree and red thread haud the witches all in dread' or in a slight variation, 'Rown-tree and red thread will put the witches to their speed'. Among the many interests of King James VI of Scotland and I of England was the practice of witchcraft, and in his study of *Daemonology*, he noted that rowan twigs were often knitted into various possessions to ward off the evil eye, a practice that persisted for many years. Rowan branches were routinely attached to mantelpieces and lintels, while children and young women would string the berries into necklaces. Farming families, their horses and even livestock would

be dressed up in decorative loops made of rowan on days when the risk from witches was thought to be most severe – those dangerous transitional days in the Celtic calendar, such as the Eve of May Day, the summer and winter solstices, the spring and autumn equinoxes.

The guardian in the garden grew up with the family, binding together the generations in peaceful security, as the popular song gratefully acknowledges:

> Oh! Rowan Tree Oh! Rowan Tree!
> Thou'lt aye be dear to me,
> Entwined thou art wi' mony ties,
> O' hame and infancy.

Caroline Oliphant's folk ballad recalls the 'hallowed thoughts' inspired by the comforting rowan tree in a devout household, where each day was punctuated by prayer and psalms. To wear a cross made of rowan was to interweave folk tradition with Christianity – a belt-and-braces approach to impending danger.

The protective power of rowan wood made it a favourite choice for cradles and walking sticks, offering extra help to the most vulnerable members of a family. Since the wood is hard and resilient, it was suitable for making masts for small sailing boats. This practical advantage was greatly enhanced by the hope that it would also ensure a safe passage for those on board. Cows were another focus of particular anxiety, so rowan was used for stirring milk and fortifying the churns against curdling. There were even fears that a dextrous witch could, by skilful manipulation of a rope made from the hair of a cow, steal the milk from other people's herds for making her own cheeses. The way to prevent this kind of behaviour was to suspend a rowan twig above the byre with red thread.

Much of the evidence for the folk traditions surrounding rowan trees survives through reports written by interested observers, often inclined to present them as quaint, faintly absurd, rustic habits that belonged firmly in the past. In his monumental survey of agricultural methods, *The Statistical Account of Scotland*, Sir John Sinclair of Ulbster noted the use of rowan for warding off elves on cattle farms near Kirkcudbright. As far as Sir John was concerned, this was an unhappy residue of the dark age of superstition, which would soon be relegated to distant folk memory by the combined forces of modern Enlightenment and practical, agricultural improvement. His desire to bring about modernisation was genuine – and it is easy to see why, in the wake of earlier witch trials, suspicions about black magic seemed so undesirable. What may be less apparent, though, is the gulf between the cultural language of older communities and that of the improvers. Traditions handed down from parent to child often go unrecorded by those who understand them best, eventually surviving only in the words of those to whom their real meanings are obscure. It is easy enough for outsiders to smile at the beliefs surrounding rowan trees or to mock them as old wives' tales, but trees can still tell us things if we are prepared to listen.

Suspicions about thefts of milk from inside the cow's udder are, after all, not so very different from worrying about having petrol syphoned out of the car overnight or finding that the satnav has been stolen. There is nothing new about the fear of nasty neighbours, or the faint sense of hidden malevolence lurking behind faces we see every day and yet do not really know. The details of what might happen under cover of darkness may change, but the feeling of being under threat from an unknown, but uncomfortably close, menace probably does not. Anxieties surrounding newborn babies, or relatives who are no longer able to look after themselves, are hardly misplaced. In communities where there were no nursery alarms or panic buttons, and no emergency services even if the worst did happen, the desire for some sense of security is easy to understand. Wholesale disasters, or just unexpected mishaps, can provoke a range of powerful emotions, but often the need to make sense of what has happened, to find some cause or explanation, is part of the overriding desire to recover control. Fear of witches may have been a way of expressing fear of the unknown or anger at the unexpected, making the rowan both a kind of early insurance policy and a witness for the prosecution. The rowan was traditionally planted as a prophylactic against all things bad, but this protective role is indicative of an acute sense of surrounding dangers and the very human urge to erect defences.

In her sympathetic study of Scottish folklore, *The Silver Bough*, Marian MacNeill recorded many of the beliefs associated with the rowan, but the most affecting passage is her reflection on the dangers that the tree was powerless to fend off – not witches, but 'a worse enemy': the factor, or land agent, whose purpose was to maximise profits from the poorest land. As the Highlands were cleared of smallholders to make room for the enormous herds of sheep farmed by supposedly enlightened landowners bent on improvement, the local people emigrated in their thousands, leaving behind their houses, byres and guardian trees. Marian MacNeill observes that 'in many a deserted glen from Skye to Angus, the rowan stands pathetically beside the roofless cottage and reddens each autumn – for shame, it might be said, at one of the darkest blots in Scottish history'.

Desire for the rowan's protection – and fear that it may not be a strong enough guard – brings this mysterious tree much closer to home. Most of us do not face precisely the same threats as the people who once lived in the now deserted cottages of rural Scotland, but everyone's world has terrors of its own. In her award-winning children's book *Whispers in the Graveyard*, Theresa Breslin tapped into the ancient beliefs surrounding the rowan tree. They still possess an alarmingly contemporary relevance once channelled through a lonely young narrator called Solomon, who has to endure daily humiliations from unsympathetic teachers, irritated by his undiagnosed dyslexia, and for whom 'home' means yearning for his absent mother and dodging the worst excesses of his alcoholic father. Solomon's only refuge is a deserted churchyard, full of quiet graves beside an old rowan tree. The narrative explores the horrors

set in motion by council workmen who arrive with an order to remove the tree. What follows draws on folk beliefs about the evils kept at bay by the rowan, but at the heart of this very modern tale is the need of a child for a safe place in a frightening world. The boy's heightened perception opens up an imaginative space for recovering a different language from the standard terms of modern life and thereby acknowledging truths that cannot otherwise be spoken. The old hope that the rowan possesses protective powers is set against an insistently modern background to reveal the terrifying experience of abandonment and youthful powerlessness.

If the desire to plant a rowan on moving to a new house strikes some as superstitious and outmoded, it is really a rational acknowledgement of the basic human desire for a place of safety. And so those ornamental trees in urban streets and on housing estates are not after all so very far removed from the wild rowans on isolated Scottish hills. Surely everyone wishes for some sort of secret retreat? A place where there is nothing threatening, where everything is nourished and lovingly tended, where children can play, where lovers can meet, where grandparents can sit in the sun. Who would want to put any of this at risk by uprooting a rowan tree?

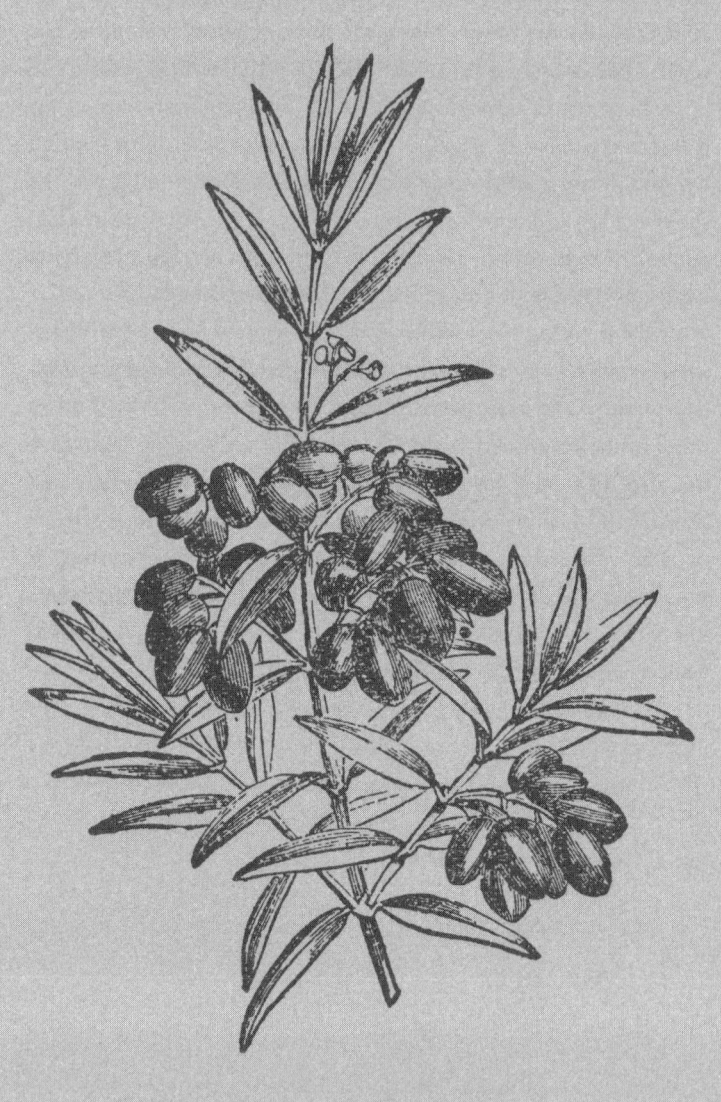

OLIVE

THE largest olive I ever saw was floating in a glass of gin and tonic. It may have been magnified by the glass and by memory, but it has certainly remained in my mind as the green standard against which all subsequent olives should be measured. The splendid, glistening, egg-like fruit was Italian, of course, and I encountered it in the ancient town of Sansepolcro in Tuscany, where I was staying for a few days for my sister's wedding. Since then, olives of every size, shape, age and origin have become commonplace in British supermarkets but, as a student in the 1980s, my experience of olives was limited to those that tended to emerge rather limply from the briny confinement of a screw-top jar. The gin-soaked, Tuscan olive was an emblem of cosmopolitan sophistication: I knew at that moment that life would never be the same again. And even now, when olives are so easily available, there is still something thrilling in their succulent flesh and subtle, uncompromising flavour, something that smacks of southern warmth and otherness.

For those who live surrounded by these slim-leaved, pewter-trunked trees, they might have come to seem mundane, but Mediterranean culture is drenched in the ancient beneficence of the olive. Fossilised olive pollen in the cauldron of the volcanic Greek island of Santorini suggests that olive trees were growing there some forty thousand years ago, though the ancestors of the domesticated olive (*Olea europaea*) may have originated in Mesopotamia. Olive trees are slow growers, but once established in suitable situations, they just keep on going. Like the tortoises that nestle in the dry grass beneath them, olives like to take their time. This is a tree capable of withstanding searing heat and water-starved soil. Where lesser plants might wilt and die, the evergreen olive drinks in the brightest sunshine, thriving in temperatures of forty degrees and above. From Spain to Syria, from Turkey to Tunisia, olive groves stud the dusty slopes with silver-green. It is the miracle tree of the Mediterranean and the Middle East, producing fruit, foliage, wood and opulent oil from the desiccated earth.

Ancient Greek civilisation grew up with the olive tree. This was the plant sacred to Athena, goddess of wisdom, as depicted in the sculpted pediment of the Parthenon. The mythic foundation of Athens depended on a single olive tree, planted by Athena, who outwitted her rival, the sea god Poseidon, by demonstrating the cornucopia of good things springing from a simple seed. The myth

RECONSTRUCTION OF THE PARTHENON PEDIMENT

testifies to the centrality of olive trees in the ancient world – providing food, wood and fuel. Greek athletes, competing in the original Olympic Games, were sleeker than the fittest Lycra-clad competitors of today, because they wore nothing more cumbersome than a coating of olive oil. The quickest and the strongest (and by then probably the sweatiest) would be crowned with a coronet woven from wild olive leaves from the *Olea oleaster* tree.

Athena's tree plays a key role in Homeric poetry, too. Throughout his epic journey home, Odysseus' progress is fraught with interruptions, but at key moments, things run more smoothly with the help of the sacred olive. The encounter with Nausicaa and her maidens, bathing in a stream, involves liberal applications of olive oil to naked flesh, though it is in the cave of Polyphemus, the gigantic, one-eyed monster, that the tree really saves the day for Odysseus and his men. To escape being eaten, Odysseus manages to seize Polyphemus' huge, olive-wood staff, heating it in the fire until it is glowing red hot, before thrusting it into the eye of the Cyclops in a graphic depiction that highlights both the benefits and misfortunes of attracting divine attention.

When Odysseus finally reaches home, he sees the long-leaved, long-lived olive at the harbour mouth, little changed from when he sailed away, twenty years before. Under the sacred tree at Ithaca, he devises a plan to recover his kingdom and his queen, and it is an old olive tree that brings about the ultimate reconciliation after Penelope's suitors have been vanquished. Odysseus only convinces his patient, but ever prudent, wife of his true identity when he shares the secret of their marital bed, created years before when he built the entire palace around an olive tree, finally lopping off its leaves, smoothing its trunk for the bed post, and finishing it with gold and silver.

The aromatic scent and the ease with which it can be turned on a lathe have always made this wood attractive to craftsmen. The extraordinary colour of olive wood, its rich brown waves rippling from pebble-shaped knots, gives it a rare liquidity. A polished slice from an elderly olive is like the secret entrance to a whirlpool, its surface perfectly still, but barely holding the energies within. Skilled sculptors, attentive to these golden waves, carve pieces gleaming with natural movement. This was the timber selected for the doors of the sanctuary in King Solomon's Temple and for the pair of cherubim guarding the inner sanctuary, the holiest chamber of all.

For ancient Mediterranean societies, the most treasured quality of the olive tree was its capacity to provide oil. When Gerard Manley Hopkins was seeking words to express God's grandeur, he described divine glory gathering to 'a greatness, like the ooze of oil / Crushed', but his imagination was soaked in classical and religious teaching. This silky, slippery, pure virginal substance was at the service of the gods of Greece and Rome, offered in libations or liberally poured over warriors and priests. The olive tree was sacred in the Islamic world, too, since the golden, translucent oil produced from its fruit was seen to reflect the divine light of Allah. This may have been inspired by the glow of sunlight shining through olive oil or by the bright flame of a burning oil lamp. In the New Testament, too, the slow, steady flame of an olive oil lamp can be a symbol for the divine or for the patient preparation of the faithful. The parable of the wise and foolish virgins tells of the young women who did – or did not – keep their lamps topped up in readiness for the arrival of Christ.

Olive oil burns very easily and brightly, as I discovered by pouring a small measure into an egg cup and standing a piece of rolled cardboard in the middle. To my surprise, the card lit instantly and, soaking up the oil, gave off a very pretty azure-based flame until, emboldened by this success, I tried holding a makeshift glass shade over it, which instantly extinguished the flame, darkened the glass and burned my fingers. In spite of this, it is easy to see why the olive tree, with its generous supply of warmth and radiance, has always been seen as such a blessing.

Olive oil, the liquid gold of the ancient world, filled the coffers of Knossos and Carthage as well as fuelling the expansion of Rome. Wherever merchants or the military went, olive trees went with

JOHN EVERETT MILLAIS, *THE WISE VIRGINS*

them (though the more northern reaches of the Roman Empire generally proved uncongenial to sun-loving groves). Olive farming is one of the greatest and most long-lived legacies of the classical world, still accounting for a considerable sector of the economy of southern Europe. The world's biggest olive producer is Spain, but Italy, Greece, Turkey and Morocco are all significant players in the world's olive stakes.

All around the Mediterranean, vast plantations of trees, organised in lines and squares, recall their Roman ancestors – though the bursting, slim leaf-heads of younger trees, all lithe and petite, tease the regular formations and seem more like a Zumba class than a military parade. Scattered amid the groves in Croatia are small, thimble-like stone buildings, with pretty sloping roofs and tiny doors. It is easy to imagine that these are the homes of hermits or wise women or perhaps some mythological creature, lost from modern memory. Known as *kazuni*, these are agricultural stores, which have not changed very much in thousands of years. According to legend, Dante slept in one on his journey through Croatia. Across the Adriatic Sea in southern Italy, the ancient olive groves surround rather more eccentric circular buildings, or *trulli*, with wizard-hat roofs, perfectly suited to the silver-headed, wrinkle-trunked trees.

The Mediterranean olive harvest, which takes place between October and December, is still a major event. Green olives are harvested first, before they have had a chance to ripen fully; those left for a few weeks longer turn dark as red wine. Olive oil is made from the ripe, deep purple fruit, which has been softened and slightly sweetened by the autumnal sun. In Andalusia, where much of Spain's olive crop is grown, the favourite variety for making oil is the taper-tipped Picual olive. In many regions, the best olives are still harvested by hand, saved from bruising by the specially designed nets and baskets. This is, of course, extremely labour intensive and may seem old fashioned, but there are practical and economic reasons for the care traditionally lavished by growers on their groves. Roman authorities advised farmers against beating the trees, not for religious reasons but because such violent haste might damage next year's crop: the slow-growing olive demanded long-term thinking in order to give of its best. Some farmers do resort to mechanical shakers, nevertheless, allowing their trees to shudder while the fruit showers down.

Olives survive well when stored in salt water, though immersion times vary around the Mediterranean, and the bitter taste of under-ripe, green olives requires pre-treatment with water and wood ash. Black olives, like the juicy Kalamata variety from Greece, are usually dowsed in vinegar or dried and packed in salt. For olive oil, the fruit has to be crushed to pulp and then pressed. This process also removes much of the bitter-tasting glucoside residing in fresh olives, as the oil separates from the watery liquid. Modern mills employ steamrollers to do the job rapidly and efficiently, but for centuries olives were crushed by great stone mill wheels, turned by a mule, walking slowly round and round the press. 'Virgin olive oil' means that no chemicals have been added during the pressing process, while 'extra virgin oil' refers to the oils with lowest acidity and highest quality.

Mediterranean cuisine is almost synonymous with the olive, because of the ubiquitous, fine-flavoured oil, which adds that very distinctive taste to dressed salads, cakes and bakes, and fried or grilled dishes. The olive fruit is very versatile, too, whether baked into bread, blended into paste for bruschetta, sprinkled over pizza, stuffed with capsicum, or launched into a cocktail. While it is not difficult to see why warm sun and stress-reducing siestas might help to prolong life, the good health associated with a Mediterranean lifestyle is probably a direct consequence of the ubiquity of the olive tree. Olive oil, a natural source of monounsaturated fat, tends to lower, rather than raise, cholesterol and blood-pressure levels, as well as being brim-full of antioxidants. The risk of heart disease, stroke and even certain kinds of cancer is reduced by a diet rich in olive oil. It is also a marvellous treatment for earwax, a quick dose clearing the passages and helping to improve hearing.

The olive tree stands for health and longevity, surrounding the blue Mediterranean with a reassuring air of continuity. Long life and quiet stability seem embodied in these familiar trees. Travellers from northern Europe, for whom the olive tree was primarily a figure from scripture or classical literature, have often been overcome by their first physical encounter. Tennyson was deeply moved by the olive trees beside the Roman ruins at Lake Garda, so little changed since the poet Catullus had described them almost two millennia before. The people were gone, their villa a wreck, but the trees at 'Sweet Catullus's all but island, olive-silvery Sermio' were as fresh as ever.

That olive trees have been growing at the same site since antiquity can sometimes be quite literally true. All around the Mediterranean there are trees a thousand years old and more. The oldest olive in Portugal, at Santa Iria de Azóia in Loures, is more than 2,700 years old, making it coeval with Hellenic Greece. In Puglia, where there are thousands of *ulivi secolari* (age-old olive trees – many twelve centuries at least), each has been carefully recorded through satellite mapping in a remarkable combination of modernity and antiquity.

The islands of Briyuni, in the Adriatic Sea, where Marshal Tito had his summer palace, are home to all kinds of surprising things. Surviving photographs show the Yugoslav president entertaining the global greats of his era, from Nehru and Nasser to Burton and Taylor. It was customary for honoured guests from far-flung parts to present a distinctive gift from their homeland, and so the island zoo includes a herd of zebra, a giant turtle, and even an elephant from Indira Gandhi, all now decidedly long in the tooth. The most distinguished veteran on the island, and certainly the one with the largest trunk, is the ancient olive tree. Local legend has it that the tree was planted by the Romans, a claim more or less confirmed by recent carbon dating, which puts its age at 1,600 years at least.

This remarkable old tree shows itself quite differently, depending on the angle of approach. From one direction it looks to be a model of strength and self-control, its stout trunk supporting a vast, gravity-defying canopy, spreading asymmetrically over what gives every appearance of being a perfectly flat platform. It is hard to see how it can possibly balance and yet, there it is, as still and serene as a vast green cloud. As you move around the tree, it is clear that the other side has not fared quite so well, with one huge branch collapsed onto the ground and half the centre split away. And yet, the broken bough has sprouted and the shoots from its vast limbs are larger than many of the olive trees you see in the fields on shore. Olives have remarkable powers of regeneration: the feathery leaves of these phoenix trees will sometimes flourish again after they have been razed to the ground.

The trunk of an ancient olive can look like a petrified river delta, spreading so broadly into the earth that it is hard to imagine how it could ever fall over. Some are so grey and smooth that they look more like boulders than tree trunks, others so twisted that they seem to have been screwed into place by the ancient gods. The olive tree's reputation in the classical world for withstanding whatever fortune might throw at it is evident from Aesop's mildly mischievous fable about the olive and the reed. Here the olive, proud of its great age and certain of its long-held opinions, is contemptuous of the reed's habit of giving in to every wind that blows. The tables are turned when a serious storm sweeps in: the adaptable reed survives unscathed by bending with each strong gust, while the poor old olive breaks up in the blast.

Since the olive's capacity for standing its ground is so well known, it is odd that even the most ancient trees are currently prone to transplantation. No fashionable villa is complete, apparently, without its own gnarled olive tree, and so unsuspecting veterans are regularly uprooted from their customary surroundings to be replanted as striking garden features. Ancient Spanish trees, over a thousand years old, have been shipped across the Atlantic or hoisted by helicopters to wealthy estates across Europe and the Middle East. Too large and heavy to move by road, such transportation of ancient olive trees does raise serious questions about whether this is a legitimate source of income for struggling Spanish farmers or an act of cultural vandalism. So valuable are these trees that there is even an illegal trade – olive kidnapping. Defenders often declare a conservationist motive and, when an ancient rural habitat is threatened by new development, olive tree transporters may be performing a rescue rather than a raid.

The removal to Antalya of Turkey's oldest olive, planted in the Aegean province of Izmir in 1071, has been decidedly controversial nevertheless. Although its journey was prompted by the international exhibition taking place in 2016 at the Botanical Centre in Antalya, the sight of the stout, spreading veteran being unceremoniously snatched from its bed and carried off by a tall, orange crane has caused widespread concern. Olive trees are amazingly adaptable, however, and with very careful excavation and skilful trimming, their hardy roots will often withstand the shock of removal and replanting. If all this seems to render Aesop's fable a little redundant, it does offer some hope that the current fashion for moving elderly trees will not prove fatal.

The olive tree's rootedness – or otherwise – can also be a matter of powerful political feeling in conflicted regions. A modern Palestinian protest song describes its people as being 'uprooted like our olive trees'. For Israelis, too, the olive tree is a symbol of their nation and the leaves feature in stylised form around the menorah, in the national emblem designed for the foundation of the modern state after the Second World War. The significance in the Old Testament of the Mount of Olives, which rises in the upland ridge to the east of Jerusalem, means that it is at once a sacred and contested site for Jews, Christians and Muslims.

Whether something of this informs a poem by the Belfast poet Ciaran Carson about the experience of living in a divided city is left characteristically understated, but in 'Patchwork' he recalls the shattered panorama of home and his father's 'rosary of olive stones from / Mount Olive, I think, that he had thumbed and fingered so much, the decades / Missed a pip or two'. Contemporary poets in Northern Ireland are painfully attuned to the symbolism of the olive branch. In his moving elegy to Charles Donnelly, the Irish poet killed in action at the age of twenty-two during the Spanish Civil War, Michael Longley imagines the young man just before he dies, picking a bunch of olives from the dust and squeezing them 'to understand the groans and screams and big abstractions / By saying quietly: "Even the olives are bleeding" '. And yet, the tree goes on growing over the grave, bearing fresh wood and fruit and branches. Like Carson's image of the incomplete olive-stone rosary, the shadow of the Spanish tree evokes, with poignant irony, the olive's ancient association with peace.

For the olive's most celebrated appearance in the cultural history of the human race is described in the Old Testament. As the waters of the great Flood eventually begin to recede, a white dove lands on Noah's Ark, bringing an olive branch. The first sign of recovery, of God's forgiveness, of a future more blessed and peaceful than the past, comes in the shape of a sprig of olive leaves. Fortunately, the tree's foliage is simple enough to recognise and so a stylised

THE DOVE AND THE OLIVE BRANCH

branch of slim, V-shaped leaves has become an international symbol of hope and harmony. This is the tree under which all the nations of the earth will come together to act in the common interest of humanity and attempt to secure that most precious state – peace.

Long before the United Nations adopted the olive branch as its logo, the tree featured as a symbol of peace in religious and political iconography. In *Antony and Cleopatra*, Shakespeare has the victorious Caesar, bringing news of 'universal peace' after the Battle of Actium, proclaim that 'the three nooked world shall bear the olive'. The same biblical metaphor is evoked by the Earl of Westmoreland after the failure of the rebellion in *Henry IV, Part 2*: 'peace puts forth her olive everywhere'. This is not just the sprig carried by the dove of peace, but an image of a flourishing olive tree that stands for health, prosperity and divine favour. Again, the precedent is biblical, for in Psalm 52, David offers thanks for his sense of enduring safety and security: 'But I am like a greene oliue tree in the house of God: I trust in the mercy of God for euer and euer.' After the English Civil War, parliamentarians drew on the ancient symbol to emphasise the peaceful associations of Cromwell's Christian name. A loyal medal was struck, featuring Oliver on one side and an olive branch on the other, while political prints poured from the presses depicting the Lord Protector surrounded by vigorous trees symbolising divinely ordained peace. These were not very welcome in areas resistant to his victory, of course: for many in Ireland Cromwell's olive branch seemed grotesquely distorted.

Gratitude for the olive of peace generally springs more spontaneously in the words of those on the winning side. James Henry Leigh Hunt, celebrating what turned out to be only a brief cessation of the conflict that overshadowed Europe for more than two decades, hailed peace as a sacred plant:

> Divinest of Olives, O, never was seen
> A bloom so enchanting, a verdure so green!

Lord Byron, writing after the final defeat of Napoleon, was rather less sanguine about the condition of the traumatised, impoverished post-war world and tended to deploy traditional symbols of reassurance more cynically. In his epic, *Don Juan*, the distinctly un-Odysseus-like hero finds himself adrift for days in a small boat, with a handful of starving, dehydrated and steadily diminishing companions. Eventually, the sight of a beautiful white bird seems a good omen, but the poem's caustic narrator quells any thoughts of providential delivery: ' 'Twas well this bird of promise did not perch,' he observes, before reminding us of just how hungry the crew are:

Had it been the dove from Noah's ark,
Returning there from her successful search,
Which in their way that moment chanced to fall,
They would have eat her, olive branch and all. (Canto II, 95)

Symbols of peace are not always quite sufficient for the desperate. Nor do they necessarily bend to the unpredictable course of history. Canova's magnificent statue of Napoleon in his prime, posing nude in the classical mode as Mars the Peacemaker with his sword hanging unused from an adjacent olive tree, acquired a certain irony after it was given to the Duke of Wellington following the latter's victory at Waterloo. The Martian model was, by then, on his long voyage into exile in St Helena, leaving Europe to cultivate the olive groves.

After his defeat, a number of Napoleon's leading supporters fled across the Atlantic in the hope of finding a safe haven in the United States. They were allowed by the US Congress to settle in Alabama on the condition that they applied themselves to the peaceful task of cultivating vines and olive trees. Unfortunately neither of these hallowed and lucrative Mediterranean plants did very well, and the colony declined along with the trees. Later in the nineteenth century, settlers discovered that central California, with its warm limestone valleys, offered a much kinder nursery for young olive

ANTONIO CANOVA, *NAPOLEON AS MARS THE PEACEMAKER*

trees, and today almost all the olives grown in the United States hail from there.

The desire to grow olives runs deep and wide. As the prospect of global warming began to turn into a reality, entrepreneurial garden centres in the United Kingdom were quick to put a positive spin on rising temperatures by encouraging their customers to opt for Mediterranean themes. Over the last decade or so, fashionable garden makeovers have routinely included arrangements of small olive trees, inserted into elegant square stone planters of Italianate design. Not all of them thrive in the November blasts, torrential rain or deep snow of some British winters, but it is probably the lack of sustained sunshine in July and August that causes the real trouble. Whether or not these slim, pompom-topped saplings live to fulfil their promise, the fact of planting them at all is testament to a heroic impulse to turn the most unlikely spot into a haven of Mediterranean abundance. The very name of the tree in English, anglicised from the Latin *oliva*, is somewhere between a command and a plea: 'O live!'

The olive glows as an emblem of hope, irrespective of historic uses, abuses and misuses. Its long, long life carries a reassuring sense of continuity, and its extraordinary powers of survival nurture hope for the future of even the most conflicted regions. This is a tree that has been growing in the Middle East since before the earliest human records and there is no real reason to doubt its capacity to continue for millennia to come. Even after an olive tree has been scorched and burnt, it may still sprout fresh shoots and begin another life.

CYPRESS

Anyone ascending into the hills above Florence will be struck by the slim, dark trees standing sentinel on the way to Fiesole. The enigmatic figures of cypress trees punctuate the lines of cars and coaches, giving pause for thought. The slow ascent from the intense streets and suffocating heat of Tuscany's Renaissance city state allows for a quiet widening of the vista, a gradually enlarging sense of a defined place held somehow within an intangible but even more ancient whole. Indecipherable Etruscan characters lie hidden beneath the green-black striations of the still, unchanging cypresses. In Italy people plant these trees to sweeten the sometimes stale air of a hot summer, but with the aromatic aura comes a sense of half-forgotten things, of sorrows long ago. The Latin name of the Italian cypress, *Cupressus sempervirens*, means always living, but this is a funereal tree.

Across Europe and the Middle East, cypress trees are planted in cemeteries to form evergreen columns between the tombs. In Japan, cypress wood is in demand for coffins and shrines. Cypress trees make suitably sober, flame-shaped attendants at Indian temples, too, while their pungent timber forms an essential part of the funeral pyres, softening the acrid smell of burning flesh. The strong perfume of cypress wood has always been unleashed to help the soul on its way, but it is now known to release a natural fungicide into the air, cleansing the atmosphere and protecting the mourners. In Britain, too, gloomy associations have tended to hang about these trees, which were widely regarded as symbols of mortality and thus immune

CYPRESS TREES IN TUSCANY

from pruning. Even today, a row of Lawson cypresses, buffeted by sharp winds in January, easily brings to mind a forlorn procession of elderly aunts, wrapped in thick coats, attending a committal.

The tough wood of the cypress tree is legendary for its durability. Plato's Laws were engraved on tablets of cypress wood, chosen for its natural longevity and gravitas. Cypress timber went into the building of the great bridge over the Euphrates and the doors of St Peter's Church in Rome. This tree was even believed to be the biblical gopher chosen for the construction of Noah's Ark, admittedly with a little etymological ingenuity. Above all, it was the timber as enduring as death itself: the bodies of Greek heroes were incarcerated in cypress wood, and it was used by the Egyptians for the chests housing mummies.

The cypress tree has always been shrouded in unhappy associations. Ovid tells a tale about the tree's former life as the beautiful boy Cyparissus, favourite of the god Apollo. Cyparissus, in turn, was devoted to a magnificent stag, which he would ride with scarlet reins, after decking its golden antlers with flowers. On one summer day, when the stag was resting in deep shade to keep out of the excessive heat, Cyparissus was practising the javelin and accidentally killed him with an unlucky throw. The boy was heartbroken, and so shocked by what he had done that his hair stood up on end. As he wept inconsolably, all the blood drained from his face until eventually he turned into a dark green tree. Apollo grieved for Cyparissus, who was doomed to weep for ever and to become the constant companion of those who mourn.

Troubled artists have often been drawn to cypress trees. In Edvard Munch's disturbing painting, *Golgotha*, with its crowd of nightmarish faces staring out at the viewer rather than towards the crucifixion, the left side is bordered by a pair of inscrutable cypress trees, standing exactly parallel to the upright cross, but seemingly offering little reassurance of the resurrection. There is nothing spring-like in Paul Nash's late painting, *Landscape of the Vernal Equinox*, either. For all its geometric order and the mysterious conjunction of sun and moon, the foreground is dominated by the dark, untwinned obelisk of a portentous cypress tree. Death has ever been present in Arcadia, and the cypress still stands as its natural representative.

In the months before his suicide, van Gogh painted *Road with Cypress Trees*, in which a huge cypress looms up behind a pair of diminutive human beings, mocking their tiny spade and jaunty stride. Neither dares look over his shoulder at the towering tree. This painting echoes the brilliant works van Gogh had created in the glorious sunshine of the previous summer in 1889, when he was

VINCENT VAN GOGH, *ROAD WITH CYPRESSES*

committed to an asylum in the south of France. In his famous paintings *The Starry Night* and *The Wheatfield*, the churning skies, thick bright stars and vibrant colours are skewered by the sinuous shapes of blackening cypress trees. Van Gogh explained to his brother Theo that a cypress tree was 'the *dark* patch in a sun-drenched landscape', but he was also compelled by the tree's capacity to sound 'one of the most interesting dark notes' in that landscape.

Cypress trees seemed to speak to van Gogh in his most creative and most tormented state. Perhaps by giving them form on his canvases, he was able to force out some of the demons from his mind.

The solemnity of the cypress tree is reinforced by its curiously sonorous character. This is the wood of choice for makers of church organs and other musical instruments. Not only is it especially resistant to fungal infestation and woodworm, its natural form lends itself to long, smooth pipes. Once the thick foliage of a mature cypress ceases to cling tightly to the bark, the exposed trunk with its upright branches looks rather like an organ, though the music of the living tree is more of a sigh than an anthem.

During the Romantic period, the cypress tree became a byword for melancholy. In his lament for Thomas Chatterton, the brilliant boy-poet thought to have taken his own life at the age of eighteen, Samuel Taylor Coleridge wove his imaginary 'cypress wreath' as he mused on the young genius wandering alone by the Avon. Percy Bysshe Shelley's 'Alastor; or, The Spirit of Solitude' portrays a young poet-hero, ranging through fantastic oriental landscapes, who succumbs to an early death in a waste wilderness so remote that there are no 'weeping flowers' or 'votive cypress wreath'. In similar vein, though with a little less verve, Bernard Barton actually addressed a poem 'To a Cypress Tree', declaring that 'Mourners love the cypress tree', while the even less immortal bard, George Darley, launched his sonnet on the same subject with the heartfelt apostrophe, 'O melancholy tree!' It was no surprise, then, that Byron's irresistible rebel without a cause, the self-exiled Childe Harold, sought relief among the cypress groves of the Mediterranean from whatever it was that kept driving him on, further and further from his native land. Even less surprising was the reference to the melancholy tree in Thomas Love Peacock's breezy contemporary satire of fashionable self-obsession, *Nightmare Abbey*. Peacock had no need to resort to a floppy quiff, curling lip and flowing cloak for his caricature of Lord Byron – he simply called his comic character 'Mr Cypress'.

THE EDINBURGH TREE TRANSPORTER

For Byron and his contemporaries, the cypress was the signa-ture tree of the Mediterranean and the Middle East, but different species of cypress grow naturally on every continent – this is one of the few species of tree to have such an international character. As European settlers pushed ever westwards across America, they discovered the Monterey cypress in Oregon and the Nootka cypress in Alaska – both larger and less tight-lipped than their Italian cousins. In the swamps of Florida and Louisiana, a magnifi-cent species of cypress rears up out of the dank water. Its roots, apparently reluctant to be drowned, send up tall, weirdly shaped growths, so the tree has its own circle of spiky defenders. Unlike every other cypress, this Goliath of the swamps is deciduous and so goes by the rather unheroic name of the bald cypress. There is still some debate over whether it can be a true cypress or is merely an impostor. On the far side of the Pacific, similar uncertainties surround the weeping cypresses of central China and the Hinoki and Sawara cypresses of Japan. In African deserts and the Australian outback, on Mexican rocks and Chilean river-banks – wherever there is a botanist, there seems to be a new species of cypress tree.

Once introduced into Britain during the nineteenth century, many different kinds of cypress from around the globe met and mated. The now ubiquitous Leyland cypress was the progeny of a Monterey cypress (*Cupressus macrocarpa*) and a Nootka (*Cupressus nootkatensis*), which arrived separately as visitors to the Powys estate in Wales and got on very well indeed. When the next heir inherited the estate and the Leyland surname, he transplanted the new hybrid to his Northumbrian estate at Haggerston Castle – and the rest is history. Victorian plant hunters and breeders were immortalised through their trees, just as surely as Cyparissus, though generally less unhappily. Several modern varieties of cypress take their name from Lawson's Nursery in Edinburgh, where an entrepreneurial Victorian market gardener became very rich by spotting the growing demand for exotic conifers.

As evergreen mania gripped nineteenth-century Britain, public parks and gardens grew. In Edinburgh, the old Botanic Garden below the Mound had to move to make way for Waverley Station. The new site at Inverleith offered far more space and fresh opportunities for expansion, though this led to some logistical challenges, as prize specimens got in the way of ambitious plans for palatial new greenhouses. Undeterred, the gardeners constructed an enormous wooden tree transporter, complete with wagon wheels and a pulley system. A surviving photograph shows a great team of men in waistcoats pulling it along, with a vast cypress tree riding high above. The scene is directed by a poker-straight bowler-hatted gentleman in a tailcoat – perhaps the curator of the Botanic Garden – and the audience is an ample (and evidently sceptical) lady with her hands on her hips. The massive cypress was undoubtedly an A-list celebrity.

Nor is this at all surprising – for a mature, flourishing cypress tree is a magnificent sight. If planted with sufficient room to expand, a cypress will grow into a gently curving spire, standing proud against an open slope or group of paler, deciduous companions. They are often bold, handsome trees, though their feathery foliage reveals a softer side. In spring, some varieties sprout masses of tiny magenta flower balls, making the dark green sprays of foliage flush with unexpected colour. Some gleam with golden-frosted fringes, others are splashed with pale blue, like a misty headland washed by the sea. During the winter, too, these trees stand up to the weather, holding their own as all the surrounding leaves succumb. In the dead month of November, the cypress tree still lives up to its Latin name – *sempervirens* – offering a warm, dry, safe haven for roosting birds and small, winter-bitten wildlife through which to survive the dark, difficult months ahead.

A tall, shapely cypress tree makes a defining feature in a grand Italianate garden, but it will also adapt to the confines of a more modest urban space, acting as a tactful screen for ugly breeze-blocks and awkward corners. Cypress trees stand shoulder to shoulder to form a solid barrier against the wind, and they are often to be spotted on golf courses, making their way across a smooth shaven slope like an enormous centipede. They can protect suburban homes from busy roads or railways, the incessant roar of traffic and noise of trains muffled by their thick green foliage. In more polluted inner-city areas, the perfume of the cypress tree helps to counter unhealthy fumes, which is why John Evelyn recommended using the timber for doors and fences in the plague-ridden streets of seventeenth-century London. (The scent is, indeed, so powerful that if you accidentally brush against a cypress tree in late September, the invisible waft of aromatic air can catch you off guard.)

The most striking feature of the cypress tree is its extraordinary capacity to put on height – it can grow as much as three or even four feet a year, which means that to anyone looking out on an enormous eyesore, the cypress tree is a friend indeed. That is unless your eyesore is your neighbour's pride and joy. One man's grow-your-own green screen may be another's advancing enemy. And a tree that can grow to thirty feet in ten years can keep on growing – to sixty, eighty, a hundred feet. The tallest Leyland cypress in Britain measured 36.5 metres (120 feet) in April 2015, and it is still growing. The oldest known *Cupressus sempervirens* is in Iran, and has been expanding for some four thousand years, so it is very difficult to predict how long the younger Leyland hybrids might last. Not everyone wants an army of giant conifers lining up next door.

During the 1990s, sales of leylandii rocketed – and so, of course, did the trees. As a result, these grand trees, with their ancient pedigree, are now frequently regarded as nuisances. Nothing, it seems, is more likely to antagonise new neighbours than setting out a row of cypress saplings before the packing crates are empty, as Nicole Kidman discovered when she had 150 planted on her new estate in Southern Australia. She had not even chosen an indigenous variety. A long-running dispute between the former head of the Royal Bank of Scotland, Fred Goodwin, and his neighbours over the 25-foot-tall row of cypress trees surrounding his shady property ended when one of them set about solving the problem with a chainsaw. The disagreement might have been settled in court, since in 2013, a new High Hedge Act was passed in Scotland. Courts do not necessarily come down on the side of the anti-cypressians, however, as a group of homeowners in Lanarkshire found when they sued their neighbour after years of anguish over an ever-increasing wall of evergreens. The final judgment was that the soaring, fifty-foot trees must be reduced to a mere twenty – a compromise doomed to please neither side.

Does the right to privacy trump the right to light? Many councils in Britain routinely issue advice on how to tackle an offending row of cypress trees (or rather their owners) and even employ high hedge officers. The idea is to resolve disputes before they reach the courts, but in 2013 the village of Barnoldswick in Lancashire was gripped by a row between neighbouring gardeners. A 39-year-old resident was charged with battery after he turned his hosepipe on his neighbour, jetting her with such force that she fell off her stepladder. In court, the defence argued that it was all purely accidental, but as the lady in question was hosed while in the act of cutting back her neighbour's very substantial leylandii hedge, his innocence was difficult to maintain beyond all reasonable doubt – though the conviction was later overturned. The Antisocial Behaviour Act of 2003 had to be rapidly amended to accommodate disputes of this kind – though it took lawyers eighteen months to agree on the legal definition of a 'high hedge' (it is two metres – and anything exceeding that is a too high hedge). The cypress tree is now the most likely candidate for an arboreal ASBO.

Cypress trees have provoked quite desperate behaviour – the first-time offence of one Lincolnshire pensioner was to inflict criminal damage on cypress trees. Every night he would secretly relieve himself on his neighbour's fine, bushy row of leylandii, with the result that they were no longer quite so fine and bushy. It was a slow, rather smelly, death for these healthy trees and would have been the perfect murder – except that he was caught on camera. What was it that provoked such silent aggression, turning a garden of all places, that haven of peace, harmony and quiet reflection, into a battleground? The cypress tree, with its dark, exotic form, is capable of launching a thousand snips – and when the secateurs are wielded by an irate neighbour, a minor dispute can quickly escalate: stealthy midnight comfort stops are as nothing to arson, vandalism or common assault.

The cypress tree is a great divider. It separates gardens and opinions. What offers privacy to some offends others. Whether it is merely that these are the quickest trees to spoil a view, or that the heavy shade and corresponding costs of the electricity needed to alleviate the gloom are the cause of such fury probably varies from hedge to hedge. Perhaps people are dismayed by the way cypress trees suck the goodness from all the surrounding soil, seeing them as the bully boys, or greedy gluttons, preventing the other plants from growing. Or is there something more profound going on?

Cypress trees may be the scapegoats of the garden – the sacrificial lambs in all the microcosmic but ubiquitous territorial disputes.

But perhaps beneath their solemn exterior lies something inherently mischievous? Do cypress trees harbour some lofty disdain for the mere human beings who shrink to smaller and smaller proportions as the years go by? Gazing up into one of these tall conifers, we may even catch a sardonic smile or a supercilious nod at our expense. If we do, it is probably a sign that the neighbours are beginning to get to us.

There *is* something about these thick, tall trees, though, that imposes on people's peace of mind. A young, slim cypress sapling rapidly swells to dominate any view. These trees seem to threaten our very sense of self as they quietly encroach. They loom in the shadows of our most unsettling dreams, inscrutable and faintly ominous. The uninvited guest at the table, the shadow cast over Arcadia, the dark note that sounds through the safety of the garden, these aromatic, eternal attendants are always there, assuming the shapes of unspoken fears, of things we dimly know but dare not acknowledge. And, in the midst of all our insecurities, these tall, imperturbable conifers stand silently by, taking on our desperate projections, but remaining largely unconcerned.

OAK

✳

MANY people have been inside a Royal Oak. It is, after all, one of the most popular pub names in Britain, outnumbered only by the Red Lion and the Crown. Unlike its rivals, though, the name of the Royal Oak is not confined to the signboard, because it permeates the very fabric of the place. Once inside, you may well find yourself leaning on a well-worn, well-polished bar or sitting on a window seat set into a wooden panelled wall, with old, ring-marked tables and an open log-fire – in other words, entirely surrounded by oak. The rows of shiny horse brasses are probably secured to an old leather harness, which was once tanned with oak bark to make it fit for work in all weathers. Decorative plates hang on the walls, displaying tender encounters under English oak trees, a hunting scene set in an oaken forest or a vignette encircled by a border of lobed leaves and acorns. If you order a pint of beer, a glass of wine or a shot of whisky, their distinctive flavours and rich colours will have been deepened by the tannin leaching from the oak barrels. The blackboard with 'Today's Specials' probably includes smoked salmon or cheese, kippers or gammon, which may all have been cured in a traditional smokehouse using the best oak sawdust. Oak is such an integral part of British culture that we hardly notice its presence. It is just there – in our homes and parks and public buildings, on our plates and under them, on medals and stamps, trademarks and car stickers. The oak is a constant presence and unobtrusive source of endless, invisible connection.

If anyone in England were asked to name the national tree, the answer would undoubtedly be the oak, though, oddly enough, this would also be true for people in Bulgaria, Croatia, Cyprus, Estonia, France, Germany, Latvia, Lithuania, Moldova, Poland, Romania, Serbia and the United States of America. Poland's founding myth is rooted in the giant oak on a hill, where a great eagle nested, inspiring Prince Lech to build his own nest – or rather kingdom – there, while his brothers, Czech and Rus, set off to establish their realms in the south and east. The most famous trees in modern Poland are the three veteran oaks named after these legendary brothers, which grow in the park at Rogalin, near Poznan, though Czech is now beginning to show his age. In Germany, oak trees were also a symbol of national strength, planted in war cemeteries to form the groves of heroes and commandeered by Bismarck as an emblem of unity. The oak appeals to separatists, too: the flag of the Basque Country depicts a shield ringed with a wreath of acorns and oak leaves. Everyone, it seems, wants to claim this tree as their own.

Sturdy, stalwart and stubborn, the oak has always been admired for its staying power. As early as the first century BC, the Roman poet Virgil singled out the oak for its enduring strength, praising the depth of its roots and its consequent ability to withstand even the severest onslaughts from the weather: 'Hence no winter storms, no blasts or rains uproot it; unmoved it abides, and many generations, many ages of men it outlives, letting them roll by while it endures.' Virgil's patron, Augustus Caesar, chose to wear the Civic Crown, the oak-leaf wreath that signified Rome's highest honour, when being immortalised in marble. In Ancient Greece, too, the oak was the tree of Zeus, most powerful of all the gods, whose commands were interpreted through the rustle of the oracular oak leaves at Dodona; and in Norse mythology, the oak was known as the tree of Thor, the thunder god.

The strength of the oak is immediately evident. Whether you come across a single tree standing straight by the gate of an ordinary field or a whole company of oaks, dotted across a vast expanse of

AUGUSTUS CAESAR WITH OAK-LEAF CROWN

grassy parkland, the sheer physical power of the tree is unmistakable. No other tree is so self-possessed, so evidently at one with the world. Unlike the beech, horse chestnut or sycamore, whose branches reach up towards the sky, the solid, craggy trunk of a mature oak spreads out, as if with open arms, to create a vast hemisphere of thick, clotted leaves. The tip of a single twiglet might sprout four of five of these lovely, irregularly rounded leaves, and as each twig sends out any number of twiglets, a whole tree might be covered in as many as 250,000 leaves. As the temperatures rise in August, oaks put on an extra layer of foliage to make up for the mayhem caused by early summer moth larvae.

The copious canopy of the oak attracts colonies of insects, birds and small animals. Tree creepers, nightingales, thrushes and wrens

AGNES MILLER PARKER, OAK BUDS

all move with relative safety, camouflaged against the rich brown trunk, but the oak also gives shelter to colourful redstarts, robins, nut hatches and wood warblers. In the older hollows, woodpeckers, little owls and barn owls will build their nests, though they might have to fend off the magpies that also frequent this densely populated tree. While blue tits and robins gobble up the caterpillars of oak moths, jays are so addicted to acorns that they will carry off as many as ten at a time, though this does give them a rather bulky fuselage. Since the oak is a favourite habitat of more insects, lichens, butterflies, beetles and fungi than any other kind of tree, it is the ideal home for birds, squirrels, dormice, bats and snakes. And that is before you begin to explore the life nurtured in the thick layers of dead leaves or the rotting heartwood of fallen branches. The oak carries an entire world within itself, but its sinewy, Atlas-like limbs show no sign of strain. This is the King of Trees, the head, heart and habitat of an entire civilisation.

In eighteenth-century Britain, the oak tree was celebrated as 'the perfect image of the manly character', because of the obvious strength of its reassuring branches, the reliable consistency of its timber and, more figuratively, its patience and good sense. The enthusiastic poet-gardener William Shenstone summed up its contemporary appeal: 'As a brave man is not suddenly either elated by prosperity, or depressed by adversity, so the oak displays not its verdure on the sun's first approach; nor drops it on his first departure.' There was nothing flighty or impulsive about these great trees, nor did they give in at the first sign of trouble.

Pride in the virility of a tree may now seem a little eccentric, but then so were the oaks – and that was a large part of their appeal. The manly oak became something of a status symbol for owners of great estates, who prided themselves not just on 'My wife, my house, my horse', but also, it seems, 'My tree'. Wealthy gentlemen were increasingly portrayed in front of their own oaks, as in Reynolds' paintings of *Lord Donoughmore* and *Master Thomas Lister*, or Gainsborough's famous depiction of *Mr and Mrs Andrews* posing under their grand tree with the estate outstretched beyond. Judging from the portrait by Joseph Wright of Sir Brook Boothby reclining in an oaken forest, the latter saw himself as a man of feeling, communing with nature; but the pose also reminded the world that he owned a lot of very valuable trees. For the pioneers of landscape design, such as Uvedale Price and Richard Payne Knight, who were owners of large estates as well as leaders of aesthetic trends, there need be no opposition between utility and beauty.

Great oaks, valuable for their timber and their distinctive forms, demanded portraits in their own right. William Gilpin directed attention to the English oak in his influential guide to the picturesque qualities of trees, *Remarks on Forest Scenery* (1791). Years as the vicar of Boldre in the New Forest enabled him to speak with even greater authority on the individual character and aesthetic appeal of old trees than he had on the picturesque tours for which he was famous. Joseph Farington's *The Oak Tree* is a portrait of one grand

THOMAS GAINSBOROUGH, *MR AND MRS ANDREWS*

specimen, resplendent in a wood in the late afternoon sunlight; in John Crome's painting of *The Porlingland Oak* the little group of bathers in the pool is put firmly in the shade by the towering tree.

In Jacob Strutt's illustrated book of Britain's grandest trees, *Sylva Britannica*, almost half the portraits are of oaks. These 'characters' were status symbols for the landowners who subscribed to the book and whose power seemed embodied in the massive, ancient trunks of their trees. The special appeal of a great oak lay not just in its striking silhouette, magnificent size and longevity, but also in its individuality. Though there were many magnificent specimens, each was unique. From the Bull Oak in Warwickshire, so called because of the bull that habitually reversed into the gnarled wooden cave to look out at the rain from under the clustering leaves, to the massive Cowthorpe Oak in Yorkshire, whose curved outline and hollow trunk inspired the design of the Eddystone Lighthouse, every tree had its own idiosyncrasies. The Greendale Oak at Welbeck in Nottinghamshire was large enough to accommodate a road, allowing the Duke of Portland's carriage to pass through its trunk, as if through a triumphal arch. The Bowthorpe Oak in Lincolnshire is so big that at one time it was fitted out with a floor and table for entertaining guests in a room that took oak panelling to the extreme.

Since a healthy oak can live for up to a thousand years, many of the trees admired by the Georgians are still alive today. The Major Oak in Sherwood Forest looks just as it did in postcards from the 1970s, the 1930s or the 1890s and, except for a large central branch, in the drawings made by the eponymous major, Hayman Rooke, a century before. When I visited the Bowthorpe Oak, it was deserted except for a few chickens taking shelter from the rain, but this only served to accentuate the scale of the vast trunk. The tree stands behind an orchard, behind a farmhouse at the end of a long lane, but if the owner is at home and kind enough to let you in, the experience is like stepping from ordinary domesticity into the presence of some immortal being – ancient, wrinkled, yet oddly welcoming.

Old oaks are individual, independent *and* inclusive. Damory's Oak in Dorset served as an alehouse, while in Bagot's Park, near

THE COWTHORPE OAK FROM JACOB STRUTT'S *SYLVA BRITANNICA*

Lichfield, homeless travellers would gather to sleep under the cover of the Beggar's Oak. The Fairlop Oak in Essex was famous for its annual fair, where every stall and puppet-show had to be pitched within the 300-foot circumference of its massive shade. Oaks are such prominent features that some have marked the boundaries between counties, providing landmarks that live on in place names such as Gospel Oak or Matlock – the meeting place by the oak. The legend of the Parliament Oak in Sherwood forest, where King John is said to have assembled an emergency meeting (a thirteenth-century version of COBRA) to discuss the Welsh threat to his kingdom, seems less strange when seen as part of a society for which trees were normal points of assembly. The Lord's Oak near Rydal Water, for example, marked the spot where the local landowner regularly called people together to discuss parish issues.

The distinctive shape of individual oak trees made them universally recognisable venues: it would not be easy to mistake the swelling figure of the Big Belly Oak in Savernake Forest. Herne the Hunter's Oak is, admittedly, less easy to identify – Shakespeare's inclusion of the tale in *The Merry Wives of Windsor* guaranteed its lasting fame, but whether the ghost with the great ragged horns was associated with a particular tree, or whether the story offered an explanation for any blasted oak in Windsor Forest, is unclear. Oaks, being tall, very well hydrated and more inclined to keep their distance than many other trees, are the kind most prone to lightning strikes. But who would want to meet at a place haunted by a chain-rattling spectral gamekeeper in any case? The tree known as Kett's Oak at Hethersett in Norfolk, on the other hand, was the rallying point for disgruntled local residents, who gathered together there in 1549 to support Robert Kett's protest against Church and King. The rebellion was rapidly crushed and Kett executed as a traitor, but the tree lived on regardless to be honoured many years later as one of the Great British Trees for Queen Elizabeth II's Golden Jubilee.

Great oaks were prized for many reasons, not least economic: timber and bark meant big business and no species was more

THE FAIRLOP OAK

valuable. Samuel Pepys, who worked at the Admiralty in seventeenth-century London, kept a close eye on the sharp practices of the timber merchants in Waltham Forest. Oak was a mainstay of the national economy, because it provided the materials for shipbuilding and therefore, trade. Oak wood was unique in its combination of hardness *and* toughness: it was partly its 'unwedgeable' quality that made people believe that lightning strikes must be signs of divine wrath, for what else could send a bolt of such power? The challenge of hammering a nail into a block of oak is still a popular game at village fêtes in some parts of Britain. This extraordinary toughness made it the perfect timber for building the strongest boats. Even slim branches were rocklike, and their decisive twists and turns supplied the 'knee-timber' from which skilled carpenters crafted the curves and brackets of enormous ships. Alexander Pope's *Windsor Forest* begins by describing a harmonious, tree-filled landscape, but throughout the poem the royal park is celebrated for its utility as much as for its history or beauty. 'Our Oaks' receive praise for carrying the rich resources of India to European markets and for

HMS *VICTORY*

providing Britain with 'future Navies'. The poem concludes with a vision of the universal peace that will follow once half the forest has rushed into the waves, as if to suggest that the trees themselves are volunteering for the merchant navy.

Although the oak's rootedness has always been so widely admired, its metamorphic capacity is equally important to its cultural meaning. At once firmly fixed and yet flexible, the oak's unparalleled strength can be mobilised, enabling it to recover a new life even after it has ceased to tower above the landscape. Something of this paradoxical quality has been harnessed by the contemporary sculptor David Nash, who has spent much of his life working with wood at his studio in North Wales. For Nash, trees suggest their own sculptures, and his recent installations at Kew Gardens include an elaborate tower of giant, balancing cups, hewn from a single oak trunk. His most famous work is the 'Wooden Boulder', which began life as part of a huge fallen oak, and then, after being shaped into a giant stone, spent years travelling gradually downstream, caught and carried by varying water levels until it eventually reached the sea. This is an oak metamorphosis, which embodies an impulse to spontaneity and movement, a desire to break free from roots and the sheltering canopy of the familiar. If the oak is at once a genial, protective home lover, it is also a fearless explorer, pushing into the unknown and at one with the wind and waves. This is the tree for islanders and adventurers as well as for those wishing to stay at home, secure within their 'wooden walls'.

Before the development of iron and steel, the oak was crucial to national defence as well as to trade and exploration. Britain's security, the very future of her children, depended on her 'wooden walls' or, in other words, the navy. Since the construction of a large vessel such as Nelson's great flagship HMS *Victory* required as many as two thousand mature oaks, trees were in great demand whenever the country was at war. Oaks are often invoked at times of profound uncertainty. An early genealogical diagram of William the Conqueror's succession, for example, shows him seated on a throne in a great oak, with his heirs descending down the trunk beneath. Ambitious rulers often promoted images of themselves as powerful hunters, charging through royal forests. The famous story of Elizabeth I learning of her accession to the throne under the great oak at Hatfield House added further weight to the royal association with oaken iconography. After the execution of Charles I, the heir to the throne was able to capitalise on his own providential escape by turning the great oak at Boscobel into a figure of legend. As the victorious Roundheads searched the woodlands after the Battle of

Worcester, the royal fugitive concealed himself within the hollow trunk of an enormous oak tree, before making his escape to France. On his eventual return to London as King Charles II, the streets were lined with people carrying oak leaves, and the date of the Restoration, the monarch's own birthday, 29 May, was declared a national holiday. Oak Apple Day is still celebrated in some villages, including Marsh Gibbon in Oxfordshire, where the annual dawn raid on a local oak for supplies of leaf sprays and oak apples is followed by a service of celebration, a parade of the village silver band, a rather extended lunch and an evening fair, all as merry as the monarch in question could have wished.

After such a precarious start to his reign, Charles II wanted to be known as a strong ruler, so Augustus Caesar's fondness for the Civic Crown was woven into his own experience at Boscobel, encouraging parallels between Imperial Rome and Restoration England. Like many a successful politician, Charles was well aware of the advantages of fusing classical authority with more personal and popular traditions. The memorable image of the king's face peering through the leaves of an oak tree, still so familiar from the ubiquitous signs for the Royal Oak pubs, derived some of its power from Augustan Rome and some from the mysterious Green Man or foliate face of medieval architecture. Monarchs whose hold on the throne is a little questionable have often turned to the oak as a symbol of continuity and ancient, masculine power. As Charles II's favourite, the oak tree enjoyed a major revival during the Restoration, fuelled by the very real need for timber. This was the moment for John Evelyn to compile his great book of trees, *Sylva*, published four years after the king's triumphant oak-lined progress and designed to stir enthusiasm for the patriotic propagation of timber trees. The steady depletion of the ancient oak forests, which had been taking place over many centuries, ultimately threatened national disaster. Oaks were already in such short supply by the Restoration period that timber was being imported – hence Pepys' concerns over the local lumber trade. The outbreak of war with Holland meant

KING CHARLES II AND THE BOSCOBEL OAK

renewed demand for ships and therefore intensive cultivation: planting an acorn was now a patriotic duty. In Ireland, too, the scarcity of oak had a devastating effect on the national tanning industry, so the call for reforestation projects was widespread. The steady disappearance of ancient trees nevertheless seemed impossible to reverse and by the following century, only 10 per cent of the land once covered by mixed forest and grassland was still wooded. The growing popularity of stories about Robin Hood and his Merry Men took place against the very real retreat of much of Sherwood Forest. The cult of the oak, hailed in David Garrick's song 'Heart of Oak', grew from fears about national security and the dawning realisation that these great trees might not be permanent after all.

The great oak cultivation project had the added advantage for British monarchs of deflecting any lingering sense of internal division outwards – towards *foreign* enemies. In the century following the Civil War, the appetite for more battles on home ground was generally small, and so, in the patriotic anthem 'Rule Britannia', the supposedly ancient, though in fact very newly created, nation of Britain is imagined rising more majestically with each 'foreign stroke' – just as the severest storms 'Serve but to root thy native oak'. This stirring piece of political rhetoric, designed to foster feelings of general Britishness, was composed by the Scottish poet James Thomson after his move to the south.

The song, which now conjures up images of waving Union Jacks at the Last Night of the Proms, was first performed in 1740, only five years before the Jacobite Rising demonstrated that shining images of national unity could not quite heal underlying divisions. The oak's strong links to the Stuart monarchy, so carefully forged by Charles II, were deeply problematic for the new Hanoverian dynasty, who had arrived from Germany and whose connection to the royal family tree was via a rather less prominent, though staunchly Protestant, branch. The challenge was to accentuate the Britishness of the oak, while downplaying or re-absorbing any Stuart associations. Britannia and her oak reached back to a mythic pre-Reformation, pre-Roman era when Britons might be imagined as a single, original native people.

As a national tree, the oak remained a contested symbol nevertheless, still featuring in Jacobite iconography, albeit more discreetly. The devastation of the Earl of Derwentwater's Cumbrian estate, sold off after he was tried and executed for his role in the Jacobite uprising, inspired images of mutilated oak trees, standing silent and reproachful, as if to question the enlightened regime that had wrought such havoc. The tree's well-established significance in Ireland is abundantly evident in the place names: Dernagree, Derragh, Derreen, Dernish, Derrybawn, Derrycoffey, Derryfubble, Derrylicka, Derrynanaff, not to mention Derry. All these places derive their names from *Doire* (or *derw*), the Irish for oak. The roll-call of famous British oaks included sites of intense national feeling for Scottish and Welsh patriots, too: the Wallace Oak at Torwood near Stirling, where William Wallace had rallied his men against the English king and the Shelton Oak at Shrewsbury, where Owen Glendower had surveyed the field. Both stood for many years as memorials to the national heroes who had once operated from within their capacious trunks.

The outspread arms of the oak offer a congenial symbol to even the deadliest opponents – and make the complicated story of its political exploitation a telling example of how different notions of nationhood can be cultivated, felled, grafted or replanted. The oak is the tree of the royals and of Robin Hood, of Britannia and Brian Boru. If the same tree inspires conservative admiration for inheritance and radical demands for equal rights, Unionist pride in inclusiveness and Separatist determination for independence, it is probably unwise to make absolute statements about what its 'true' meaning might be. The celebrated 'toughness' of oak wood, after all, embraces a sense of its flexibility and capacity for survival, which are just as important as its hardness.

The well-known longevity of the tree also offers a strong sense of hope for the future. Whatever mortal powers may come and go, it seems to say, the oak carries on regardless. The ancient oak in the park or outside the pub has always been there – and therefore it always will be. Many villages boast their own 'thousand-year-old' tree, but some oaks make these veterans seem arboreal infants. In places where peat bogs have remained undisturbed, fossilised oaks are sometimes discovered dating back to the seventh century BC. They are no longer alive, of course, but the wood of the bog oak offers a tangible connection to a world before nation states (as well as an invaluable source of prehistoric tree rings for dendrochrono-logical dating). The dark richness and strange wavy formations of bog oak come into sharp focus when transformed into striking pieces of polished modern sculpture. In Ireland, art forms created from oak spars that predate the arrival of the Celts can now be acquired for the living room. Recently, a huge section of oak that had lain submerged for thousands of years in a riverbed in Croatia was carefully raised and brought to Britain for a highly skilled team of specialist craftsmen to turn into a magnificent, semicircular bar, at once both ancient and at the cutting edge of contemporary design.

Not all oak survives quite so well. Despite its reputation for durability, oak wood has, at times, demonstrated great vulnerability to decay. The HMS *Victory*, now so well preserved in the dockyard at Portsmouth that it is easy to assume it remains unchanged since Trafalgar, nearly sank because of rotting timbers and death-watch beetle, and was rescued only by the development of new insecti-cides. The ship is a monument to an earlier era of national triumph and a reminder that the oak is not always as invincible as we might like to believe.

In his intimate account of modern Connemara, Tim Robinson registers deep emotion at the sight of a mature oak riven in two by lightning,

one half still upright but gaunt and dead, the other resting several massive elbows on the ground, living, and each half so full of its own histories, places and denizens the event must have been as momentous as Byzantium's falling away from the Roman Empire.

At least a lightning strike is commensurate with the grandeur of this tree, whose cataclysmic fall is akin to that of a great hero or even an empire.

Much less easy to reconcile with the king of the forest is – caused partly by his significant decline in the modern era – the disinclination of oak seedlings to grow in woodlands. This strange problem, identified by Oliver Rackham as the 'oak change', is a puzzling development, evident since the early twentieth century and probably caused by fungal disease or mildew, which may increase the oakling's need for light. Far from being protective, it seems, the heavy canopy of a mature oak can be fatal to tiny seedlings struggling up through the rotting, leftover leaves below.

Recently, the oak has been threatened by a series of devastating arboreal afflictions. Reports of 'sudden oak death', the fungal infection *Phytophthora ramorum*, attacking America's oak population caused international alarm and the first cases in the UK were greeted by apocalyptic news stories. Ten years on, it appears that most mature English oaks are, fortunately, resistant to the deadly pathogen, though it still tolls the death knell for many holm oaks and turkey oaks, as well as larches, chestnuts and beech trees. The oak processionary moth arrived in Britain from Europe in 2005 and since then battalions of hairy caterpillars have been marching and munching in long lines up and down the branches of their favourite trees. More worrying for the English oak (*Quercus robur*) is the disease known as 'acute oak decline' (characterised by distressing, oozing, black lesions up and down the trunk), which, after its first appearance some thirty years ago, has been spreading steadily across the Midlands and southern areas of Britain. Once the sinister striations appear, a healthy tree that has taken two or three centuries to reach full maturity may only have four or five years left to live. In the light of these afflictions, the figures of Britain's ancient oaks take on a new dimension. The Major Oak in Sherwood Forest, with its collapsing branches so carefully cradled by strongly supportive poles and wires, can be seen as a heartening image of a caring community that looks after those in need and shows proper respect to the elderly. It may also provoke less cheering ideas of a people clinging to memories of a glorious, but vanishing and ultimately doomed, heritage – a greatness that belongs to the past and will gradually be forgotten.

The very strength and longevity of oaks can make their final demise intensely lowering to the spirits. William Cowper encompassed the entire trajectory of his favourite tree in his moving poem 'Yardley Oak'. He imagines the tree's chance survival from the 'cup and ball' acorn, to the tiny twin-leaved seedling, to majestic 'King of the woods', before giving an eye-witness account of its current condition as a 'cave / For owls to roost in'. This giant of the woods might once have been a shipwright's 'darling treasure', but it was now nothing more than 'a scoop'd rind', or 'huge throat calling to the clouds for drink'. Although Cowper asserts that the Yardley Oak is still 'magnificent in decay', more affecting are the details of storms tearing off the veteran's 'arms' and leaving only stumps.

The remarkable transformation from tiny acorn to mighty oak is an important aspect of this tree's complicated appeal; its very existence is often a victory against the odds. Modern plant diseases only add to the perennial problems besetting hapless acorns, in the shape of hungry pigs, mice or squirrels, and the dangers posed to seedlings by rabbits or the gypsy moth caterpillars that abseil down from the parent tree. The undulating outlines of oak leaves are irresistible to so many species that the growth of any mature oak is quite a triumph. On the other hand, there can be benefits to possessing attractive seeds. The jay's predilection for acorns is very advantageous, since the greedy birds tend to carry off far more than they can digest and so bury spare acorns for future feasts. They return to their underground stores as much as a year later and as they tug at the shoots of the germinating seed, inadvertently help them to reach the surface. It is an unlikely partnership, but

THE YARDLEY OAK

mutually sustaining. The rotting heart of ancient oaks may have struck horror into the sensitive soul of William Cowper, but it is now plain that the nutrients in the disintegrating heartwood enrich the soil and enable the oak to regenerate. This is one of the many reasons why the well-meaning efforts of earlier generations to save ancient oaks by filling their hollows with concrete were misplaced.

The anthropological impulses stirred by veteran trees often require careful handling, since they can encourage resistance to the very methods most likely to secure the healthy future of a species. Among the more imaginative initiatives in contemporary tree culture has been the transformation of a single oak into a far-reaching educational resource. Gabriel Hemery's 'One Oak' project began with the felling of a 222-year-old oak tree on the Blenheim Estate in January 2009, an event witnessed by a large group of children. What followed was designed to demonstrate just how important an oak tree can be to today's society. The children who saw the tree fall to the ground have since gone back to collect acorns and plant the oaks of a future generation. The seedlings sprout very satisfyingly, with their young blonde leaves springing evenly from the supple central stem, like a diminutive carousel. Meanwhile, wood from the single oak was distributed among numerous crafts-people and manufacturers to be made into furniture, jewellery, musical instruments, benches, beams, doorframes, clocks, printing blocks, charcoal, firewood, sawdust and woodchips for central heating. The project may be quite different from the traditional ideas of family trees, local gathering places or national symbols, but

it shows the oak's undiminished power to draw together the disparate and disconnected.

The oak is not just a heritage tree – with proper management, this species could be essential to the future of many nations. The Independent Panel on Forestry, commissioned to research the state of Britain's trees, recently concluded that woods have the potential to make an enormous contribution to the creation of a sustainable economy. They have a tremendous capacity to lock up carbon, stabilise soil, protect areas subject to frequent flooding, improve air and water quality and provide clean, renewable energy, as well as providing invaluable nature reserves and vast, recreational areas. And in this vision of a green and pleasant future, the oak has a central role to play. A single oak is a playground and an entire natural community in itself. While its wood can be turned to numerous uses, timber joists, once proofed against disease, are strong enough to replace steel girders. A drink in the Royal Oak, with all its richly grained tables and low-hanging beams, may not after all offer a momentary step back into the past, but rather a glimpse of the future.

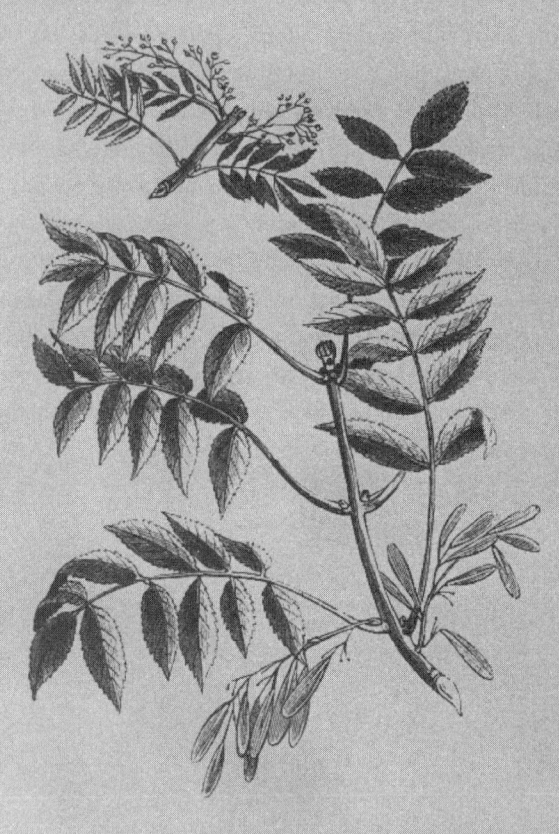

ASH

As a small child, my mother was taken to the Lake District, in the hope that she would have a better chance of survival under the shelter of the north-western hills than at home on the flat, over-exposed coast of Lincolnshire. It was June 1940. It would all have been a great adventure, were it not for the constant reminders that things were not as they should be. It was not just the absence of fathers, uncles, brothers, but the presence in the hotel grounds of oddly damaged things: a blind cat, a broken wheelbarrow, a man who had been at Dunkirk and did not seem quite like other grown ups. What my mother remembers most vividly is a young woman, pale in face and dress, who spent her days sitting outside, staring up into the branches of the tall ash tree and drawing what she saw. When the sun came out, her pencil lines darkened, turning the tracery of tiny branches into black lace veils. She never spoke, but day after day she looked up and recreated the impossible patterns on her large, flat sheets of paper. What did the ash tree mean to that unknown woman? Or to my mother, in whose agitated, impressionable mind it took root and has remained ever since?

The ash tree is known as the Venus of the woods and it seems to stir powerful feelings in those who gaze on its graceful form. Whether it is standing in spacious parkland or in an unkempt, November hedge, or rising naked from a sea of bluebells, the ash's curvy limbs taper to an end with tips pointing to the heavens. A young ash is often like a half-open peacock's tail, not quite ready to display its beauties; the branches of a mature ash, once fully fanned out, will slope down towards the earth, before sweeping up again, as if to send the buds flying. Through the summer the boughs cascade in all directions, wave-shaped and covered in green sprays. There are no angles on a young ash tree – everything is rounded and covered in fluttering foliage, soft as the feathers in a boa or the fur of a chinchilla. The boughs gain a few inches and furrows with the passing years, but with maturity come striking attitudes. In winter their silhouettes stencil clear skies like a row of unframed stained-glass windows. The ebullient black buds stand proud, as if impatient for the spring, but in fact the ash is usually the last to come into leaf and the first to shed its seasonal foliage. The uncovered form of the ash, though, is just as compelling as the full-dress splendour of more eye-catching trees.

The grace of the ash tree has always appealed to artists. John Constable immortalised the trees around his home at Dedham in Essex. In paintings such as *The Cornfield* and *Flatford Mill*, ashes are predominant in the foreground, their feathery leaves highlighted by tiny brushstrokes. According to his close friend and biographer C. R. Leslie, Constable would gaze on almost any tree 'with an ecstacy of delight', but his real favourite was the ash. Leslie recalls Constable's profound distress over the felling of an ash tree in Hampstead. The ash had inspired one of his most beautiful drawings, but Constable announced in a public lecture that 'she died of a broken heart'. He pointed accusingly to a parish notice forbidding vagrancy that had been nailed unceremoniously to the trunk as the cause of his beloved ash's death. 'The tree seemed to have felt the disgrace,' he told his audience, for almost as soon as the notice went up, some of the top branches withered. Within a year or so, the entire tree had become paralysed, and so this 'beautiful creature was cut down to a stump, just high enough to hold the board'. Constable's indignation over the fate of 'this young lady' shows that his delicate drawing was not a mere nature study: it was an expression of love.

Those looking for love have traditionally found hope in the leaves of the ash, heartened perhaps by the perfectly matched pairs

JOHN CONSTABLE, *FLATFORD MILL* (DETAIL)

of slim green leaflets unfurling from each stalk. Young women would carry a sprig of these pretty, pinnate leaves in the belief that the very next man they met would turn out to be their future spouse: a rather risky approach to speed-dating, but indicative of the ash's romantic aura. Sometimes an ash leaf or two would be tucked into the cleavage, which may have helped attract the attention of the unsuspecting swain. If the leaves were not having quite the desired effect, there was always the fruit of the ash tree, hanging down in tempting clusters and easily reached from the ground. These bunches of ash-keys were boiled to create an aphrodisiac.

CUTTING THE ASHEN FAGGOTS, DEVONSHIRE 1854

In the West Country, 'ashen faggots' raised the temperature during the festive season. On Christmas Eve, huge bundles of ash poles bound together with green ash bands to form the ashen faggot would be paraded indoors, before being laid out in the largest hearth and set alight. Those in the party would gather round to choose one of the ash bands and then watch to see which of them would catch fire most rapidly. Whoever had chosen the band that burst into flame soonest would be the first to marry. This was a drinking game, too, because whenever any of the bands burst, it was cider all round, until the whole bundle of sticks, and probably the assembled company, had become a little looser and very much warmer.

Ash trees offer promises of future bliss, but love is often mixed with less happy feelings. In his winter poem 'Neutral Tones', Hardy's disappointment in love registers in the grey, fallen leaves of the ash tree that witnesses his last rendezvous. The lesson 'that love deceives' is etched in the memory of 'Your face, and the God-curst sun, and a tree, / And a pond edged with grayish leaves'. The ash tree also stands as a reminder of lost love in the well-known Welsh folk song 'The Ash Grove':

Down yonder green valley, where streamlets meander,
When twilight is fading I pensively rove
Or at the bright noontide in solitude wander,
Amid the dark shades of the lonely ash grove;
'Twas there, while the blackbird was cheerfully singing,
I first met that dear one, the joy of my heart!
Around us for gladness the bluebells were ringing,
Ah! then little thought I how soon we should part.

There are various versions of the song, by different lyricists and artists, but each tells of love living only in memory. 'The Ash Grove' is remembered, in turn, in a poem by Edward Thomas, one of the many casualties of the First World War. Like Constable, Thomas' favourite tree was the ash, which had taken such strong root in his mind that even when he found himself in the least promising circumstances, an encounter with ash trees was charged with restorative power. The poem begins with the melancholy sight of dead and dying trees, but, in spite of this, the poet recalls, 'they welcomed me; I was glad without cause and delayed.' The trees are not those he knew before, but because they are also ashes, this grove, moribund though it may be, 'can bring the same tranquillity' as those of his past. Though hardly more alive than the dead trees, the poet experiences something of an epiphany as he wanders among the ashes:

I wander a ghost
With a ghostly gladness, as if I heard a girl sing
The song of the Ash Grove soft as love uncrossed,
And then in a crowd or in distance it were lost.

Everything is tentative and understated, but the faint stirrings gradually grow stronger, until the poem ends with a moment when the past, unwilling to die, floods the present with sudden light and an unprepossessing clump of stricken trees becomes magnified into something extraordinary. For an instant, love is no longer lost, but found, uncrossed, among the ashes.

The softness of the ash has always marked it out as a tree of comfort and healing. Wordsworth had fond memories of the mature ash at the cottage near Hawkshead where he was sent to live as a schoolboy, following the death of his parents. In *The Prelude*, he recalled how he had lain awake

> on breezy nights, to watch
> The moon in splendour couched among the leaves
> Of a tall Ash, that near our Cottage stood,
> And watched her with fixed eyes, while to and fro
> In the dark summit of the moving Tree
> She rocked with every impulse of the wind.

Many turned to the healing ash for more practical remedies. Despite the ash tree's serpentine branches, it was known to be the enemy of snakes. The Roman natural history writer Pliny the Elder, observing the antipathy of snakes to these trees (so extreme that they would not even venture under their shadow), recommended ash leaves as an antidote to snakebites. He even conducted an experiment to prove the point, placing a viper beside a small fire within a ring of ash leaves and concluding that the snake would sooner run into the flames than into leaves of this kind. Nicholas Culpeper, many centuries later, endorsed Pliny's advice on snakebites, but also recommended ash leaves for more common complaints, 'to abate the greatness of those that are too gross or fat', as well as for dropsy and gout. When infused into white wine, they became a treatment for jaundice or kidney stones. The bark of the tree, too, was used as a tonic for the liver and spleen or arthritis. Even warts could be cured by a pinprick, as long as the pin was then stuck into an ash tree.

In his *Natural History of Selborne*, Gilbert White, unconvinced by many of the traditional remedies practised in his parish, described with some dismay the row of ash trees in a nearby farmyard, whose trunks were disfigured by huge, jagged scars. Years before, their young flexible trunks had been split and then forced apart with wedges, so that naked children, suffering from 'ruptures', or hernias, could be passed through the gap in each tree. The ashes were then plastered with clay and bound tightly together again. The belief was that if the incision in the trunk healed over and the tree grew whole once more, the children would be cured. Although the physiological basis of this is undoubtedly shaky, the psychological urge to heal children through displacement is understandable enough, even if the poor little patients probably found the cure as upsetting as the complaint.

Gilbert White, as a serious naturalist and well-educated man of the Enlightenment, was troubled by the superstitions surrounding the healing ash tree. His disquiet is most obvious in his description of the 'very old grotesque hollow pollard-ash', which grew next to the church in Selborne and was used to treat sick cattle. This remedy required the ceremonial incarceration of a live shrew in the trunk of an ash tree, which was then believed to have the power to produce therapeutic twigs for stroking over ailing cows. For White, this old practice meant a triple misfortune – for the hapless shrew, for the mutilated tree and for the minds of those responsible. How could his parishioners continue to hold to such beliefs, especially when the cattle showed no sign of improvement as a result?

The kind, motherly ash was so much part of local communities that people turned almost instinctively to this tree. Numerous place names attest to their familiar presence. From Ashford in Kent to Askham in Cumbria, from Ashley in Devon to Ashwellthorpe in Norfolk, early settlers in Britain identified their homes through their favourite trees. Ashendon, a tiny village balancing on a Buckinghamshire hillside, was especially congenial to this species, judging by its name, which means 'Overgrown with Ashes'. The adaptable ash embraced successive waves of settlement, for although the word 'ash' derives from an Anglo-Saxon root, the 'by' suffix, common in northern counties, arrived only with the Danes. 'Ashby', which means a farmstead where ash trees grew, is a hybrid, but healthy enough to sustain more linguistic graftings: Ashby de la Zouche in Leicestershire reflects the influence of the Norman Conquest, while in the neighbouring county of Lincolnshire, the Latinate Ashby Puerorum was named after a bequest to the choir-boys of Lincoln Cathedral.

Wherever people lived and worked, the ash tree was a constant companion and helpmate. Its special place in men's hearts rested not only on its physical beauty or medicinal value, but, as is so often the case, on its ready supply of very versatile timber. The ash tree is probably the most adaptable resource of all for those working with wood. Its toughness and peculiar elasticity mean that ash wood can be shaped into not only relatively straight and simple constructions such as sledges and skis, but also into the least plank-like objects, including shepherd's crooks, walking sticks, bentwood chairs and even wagon wheels. The shapeliness of ash branches lent itself to such items, while the wood's flexibility also allowed for steam treatments to accentuate the natural curves.

The ash tree's readiness to furnish walking sticks has consolidated its role in the family. In Seamus Heaney's poem 'The Ash Plant', his father's stick is the 'phantom limb', steadying the old man's wasting hand and enabling him to 'stand his ground' once again. He wields it 'like a silver bough', the traditional talisman required for passage to the Celtic Otherworld. Behind Heaney's own late father is his literary forbear, James Joyce, who appears 'straight as a rush upon his ashplant' in the soul-searching internal pilgrimage of *Station Island*. Joyce's stick is not so much a prop as a baton for directing the younger poet, as he uses it to hit a litter basket before urging Heaney to 'write for the joy of it' and strike his own note.

If the ash is the Venus of the woods, it has lived up to the world of the classical deities in its age-old capacity to equip warriors for combat. Strong and adaptable, the ash tree gave its name to the spear, or *aesc* in Old English, because the straight growth of ash saplings fitted them so well for the battlefield. Ash trees have continued to arm the heroes of less deadly arenas in the form of cricket stumps, hockey sticks, billiard cues and hurleys, perhaps inspiring victory as of old, perhaps just offering the reassurance of a comfortable old stick.

With the development of iron and steel, the role of trees in national defence generally diminished and so the proud 'wooden walls' of Britain's navy were relegated to the history books. During the Second World War, however, stocks of mineral resources began to run low, threatening the production of armaments and military vehicles. In response to the crisis, Geoffrey de Havilland designed a wooden plane and, in spite of raised eyebrows at the War Office, his new Mosquito bomber swiftly went into production. Despite rationing and short supplies of pretty much everything, the ash trees of Britain remained as plentiful as ever and the skills of local furniture makers were now marshalled to make a decisive contribution to the war effort. The Mosquito bomber, light, fast and powered by twin engines,

ASH HOCKEY STICKS AND CRICKET STUMPS

MOSQUITO BOMBER

rolled off the production line in 1941, achieving fame in the raid on Oslo the following year, and playing a crucial role as a pathfinder for British Bomber Command throughout the rest of the war. In one spectacular operation, Mosquito pilots undertook a devastating attack on the broadcasting headquarters in Berlin, leading Göring to fulminate against the speed and efficiency of bombers built in piano factories. Ash had come a long way from the battlefields of the Anglo Saxons, but the same qualities of the wood that had made it suitable for spears were also adaptable for strong, lightweight aircraft.

Springy, shock-absorbent ash wood could be turned into anything from ladders and rakes to planes and even cars. It was the resilient flexibility of ash that led to its deployment in the distinctive, pale, criss-cross, boxlike frame of the Morris Traveller, as well as on round-nosed, square-backed cattle trucks and tradesman's vans of the 1940s. Ash wood is still used by the Morgan Motor Company, where modern vacuum methods have been developed in tandem with traditional wood presses to create light, elegant, but immensely robust sports cars. The Morgan production plant is full of smooth ash frames, which are ready to be covered with aluminium panels and fitted onto the strong, steel chassis. Man's very long-running love affair with the ash tree is still going strong.

Unlike some old, deciduous forest trees, such as the oak, ashes are not renowned for their longevity, most surviving no longer than a couple of hundred years. When regularly sliced in a coppice woodland, however, an ash tree will continue to send up tall, straight poles of living green, even when its heartwood has completely rotted away: at the Bradfield woods in Suffolk, the stools of coppiced ash, spreading from the ground like a great broom head, are thought to be a thousand years old. The ash's abundant keys lead to such rapid propagation that there hardly seems a need to preserve the older trees, because there is always an abundance of ashlings. Since the ash is tolerant of almost any kind of soil, it springs up all over as one of the most familiar trees in Britain. Until now.

The news that *Chalara fraxinea*, the fungus that has proved fatal to so much of Europe's ash population, is now making inroads into Britain is chilling indeed. 'Ash dieback' has already devastated woodlands in Denmark, Poland, Sweden, Germany, the Netherlands, Austria and the Baltic states, and looks alarmingly set on sweeping through the British Isles. There is something deeply sinister about

a mysterious fungus that singles out a particular species and kills off its young. In the last book he completed before his death in 2014, the veteran botanist Oliver Rackham cast a cool eye over the panic reports on the likely effects of ash dieback, observing that by the time people have noticed the presence of a deadly plant disease it is too late to take action: 'the latest year in which to react to *Chalara* was 1995'. Defence against plant pathogens requires forward thinking, he argues, pointing out in a far from reassuring conclusion that the tardy response to *Chalara* probably did not matter much in any case, because a far more devastating threat was approaching in the shape of the emerald ash borer beetle. This bright green bark beetle, *Agrilus planipennis*, native to eastern Asia, has already destroyed ash woods across the United States and Siberia, so in the era of globalisation, its arrival in Britain is almost unavoidable – unless, in an irony of natural history, the aftermath of *Chalara* provides a bleak cordon sanitaire.

Now that ash dieback has been identified and the dangers of the emerald ash borer recognised, we can still hope that the strenuous efforts to contain these threats will prove effective, that probable cases will be spotted and reported, that some resistant trees will survive and that, instead of a phoenix, the ash itself will rise again. For it is deeply distressing to be faced with the very real prospect of a future in which so many familiar wood-lined roads, green parks and sheltered towns will be left, bereft and bare, while their once perennial companions are known only from paintings, poems and old songs.

POPLAR

WHEN my children were small, the daily trip to nursery would have seemed endless, had it not been for certain landmarks along the way. Half an hour's journey for someone aged three or four, firmly strapped into a sensible car seat, seems like a whole day of hideous restraint. Lucky, then, that our journey was punctuated by familiar sights and sensations, from the roly-poly road, where an unsuccessful drainage system had resulted in a quite dramatically undulating surface, to the row of white Victorian railings with their heart-shaped tops, christened by my daughter the 'Bone Gate'. The undisputed highlight of the journey was always the long row of tall, Lombardy poplars, which contrasted so strikingly with all the bushier green trees and low-lying hedgerows along the rest of the route. Whoever planted this elegant avenue would probably have been surprised to hear the laughter it provoked each day – and the chorus: 'Mr Skinny's Trees!'

The sight of these slim, matching trees with their uptight, upwardly mobile branches seemed hilarious to children growing up in rural Oxfordshire, unlike their French, Belgian or Italian counterparts, for whom rows of poplars must be much less exciting. Lombardy poplars (*Populus nigra* 'Italica') have been common enough in the United Kingdom since the Second World War, when government planting grants enhanced poplar popularity, but they rarely line the roads like the poplars of northern France, standing tall and steady as a row of soldiers on parade. Some roads in France are still known as 'routes Napoleon', because the poplars were planted

GWEN RAVERAT, *POPLARS IN FRANCE*

under the edict of Napoleon to provide shade for troops marching in heavy uniforms under the hot, bright sun. They certainly formed a suitably triumphal avenue for the emperor – even, orderly and under control. Nowadays, tired drivers, hurrying for a ferry, can find the quick inexorable succession of tall trees hypnotic – stroboscopic, even. And yet there is also something soothing about that sense of well-regulated sequencing, the changeless pattern beside the road, striping light skies for mile after mile after mile.

Perhaps this ordered symmetry was what made these trees so appealing to eighteenth-century British landowners on the Grand Tour, who admired them in their native habitat, sweeping grandly across the Lombardy plain and along the banks of the River Po. So admirable were these elegant Italian trees that the grand tourists decided to start planting poplars of their own. The 4th Earl of Rochford returned from Rome in 1754 with the first Lombardy sapling strapped to his carriage, ready for establishing at his Essex estate. The poplar's natural vigour meant that new avenues of column-like trees shot up in no time to complement the Palladian mansions, with their grand neoclassical façades in perfect proportion. These trees seemed specially designed to match the ladies' tall ostrich-feather hats, and formed a soothing green Acropolis for gentlemen to contemplate the size of their estates. If the tree's Italian heritage appealed to fashion-conscious Britons, its alternative name was not quite so elegant, for it was widely known as the 'Po poplar'.

Rather more worrying to many of the great landowners was the poplar's role as the people's tree. Apparently springing from the same etymological roots as the populace, the *Populus* took on a strongly democratic character during the French Revolution and, as such, began to seem decidedly threatening to the nobility. Contemporary political debate over the 'natural' order of human affairs drew on images of trees, and while those alarmed by the deracination of the French *ancien régime* turned to the British oak as a symbol of slow growth, gradual change and steady continuity, radical writers such as Thomas Paine called on his fellow men to join him in raising the Tree of Liberty. The original American Liberty Tree was an old elm in Boston, but once trees became political metaphors, other species could be co-opted to the cause. Since the Lombardy poplar was strongly associated with the philosopher Jean Jacques Rousseau, whose theories about the natural freedom and equality of mankind fed the ideals of the revolutionaries, it was an obvious choice for symbolic planting. After his death in 1778, Rousseau was buried on the Île des Peupliers in the garden at Ermenonville, and so his tomb, encircled by tall, slim, matching poplars, became an icon of liberty, equality and fraternity. After the storming of the Bastille in 1789, popular prints began to circulate showing the triumphant revolutionaries planting a tall, pole-like, Liberty Tree as a symbol of the new Republic. Suddenly, setting a poplar meant planting a Liberty Tree.

Although the poplar seems one of the easiest trees to pick out in a landscape, it is worth remembering that the slim cigar silhouette belongs only to the Lombardy variety and is a relative newcomer to Britain. Before its arrival, and indeed for many decades afterwards, the majority of British people thought of poplars as large, deciduous trees, crowned with vast, delicate canopies of fluttering silver-green.

THE LIBERTY TREE

These were the natives that grew freely beside the local rivers, glowing like great lacy lanterns against the sinking sun – just as they do today, though they are generally recognised less readily than their Italian relatives. The poplar family is, in fact, bewilderingly extensive and the different species are not always easy to identify. The white poplar, *Populus alba*, also known as the abele, is probably the easiest to spot, because of its pale grey, diamond-studded trunk and the strange, white down of the leaves, which makes their under-side feel like suede. In mixed woodland, the white poplar will flash, as if moon-blanched, at the height of a summer day. These were the trees that classical authors imagined growing on the banks of the Styx: a gently shimmering entrance to the Underworld. They can now be found giving quite ordinary roads in Britain an air of mysterious grandeur.

The grey poplar, *Populus canescens*, is similar in shape to its white relation, but larger and more abundant. This species is a hybrid of the white poplar and the aspen (*Populus tremula*), so their leaves, smooth green and silver-haired, bear a family resemblance to both. The grey poplar will live for over two hundred years and can attain a height of over forty metres, if left to its own devices. At Birr Castle in County Offaly, the biggest grey poplar in these islands was, for many years, the centrepiece of the old castle grounds and thus seemed an excellent choice for representing Ireland in the European Tree of the Year contest, 2014. On 13 February 2014, however, a massive storm blew it down. The sight of the old poplar, outstretched on the riverbank, was like seeing a great leader laid out in state for people to pay their last respects. It is strange how the meaning of a tree can change utterly in a single night: from being a figure of tremendous strength and stability to one of heroic defeat, pathos and vulnerability – little more, indeed, than a pile of firewood.

Black poplars, too, were once common in Britain and across central and southern Europe, flourishing in flood plains and in the boggy, low-lying areas beside slow-moving rivers. The rapid post-war expansion of urban areas and reclamation of older, water-logged sites meant that much of their natural habitat disappeared, because the seeds germinate in bare, open ground, lying in the moist mud from June to October. Since the species is dioecious, pollination requires the presence of both male and female trees. Once they had begun to decline, the chances of natural recovery were therefore low. Oddly enough, the ebullient, easy-growing Lombardy is a kind of black poplar, too, but much more resilient to a changing, modern environment. The great black poplar, with its arching lower branches, is now Britain's most endangered timber tree – so rare that when developers in Newcastle were planning to clear the ground at St Willibrord and All Saints in the city centre, plans were halted once the old trees in the churchyard were identified as an endangered species by the National Recorder of Poplars. The great trees now lend a somewhat melancholy air as they tower over the graves on the hill above the wharf, beside the grand, disused, baroque church. One of the fifty Great British Trees of the Golden Jubilee was the black poplar at World's End Wood in Essex, chosen partly because of its rarity. And what more suitable tree for the World's End? Neither apoplectic nor apocalyptic, the black poplar seems a tree in atrophy.

The leaves of the native black poplar are not as soft as some species, but are clearly defined by their strong heart shape. This may have been what Tennyson had in mind when he depicted the abandoned bride, Mariana, marooned in a moated grange amidst the dark fen, with only a solitary poplar, 'all silver-green with gnarled bark', to mark the level waste for miles around. Poplars lose their smooth skin as they age, their trunks becoming pock-marked and deeply wrinkled, but, most poignant of all, the inevitable, seasonal fall of the native poplar resembles a cascade of faded hearts.

Although the black poplar population of Europe is threatened with irreversible decline and ultimate extinction, there are still signs of hope. In 2010, the Crown Estate launched a project to reverse the drastic decline of the black poplar in Britain by replanting slips from the Dunster estate in Somerset, one of the few remaining bastions of this native tree. This rare species also has its independent champions. Roger Jefcoate, 'the phantom tree-planter', is committed to a one-man campaign of repopulation – and spends his time combing the countryside for places to plant black poplar saplings.

He even succeeded in getting one to grow in the middle of a roundabout in Milton Keynes. His inspiration came from an earlier poplar specialist, the pioneering conservationist Edgar Milne-Redhead, who first drew attention to the plight of the native black poplar after conducting an extensive survey of their distribution from 1973 to 1988. As a botanist at Kew, Milne-Redhead's research led him to start propagating poplar trees from cuttings and popular support from essays and broadcasts. Tree survival tends to depend ultimately on a collective effort, though it takes the special insight of individuals to recognise what is happening, especially when the trees in question are so familiar as to be virtually invisible to those beneath.

The future of Britain's other native poplar, the aspen, is more secure, as these trees can be found across much of the northern hemisphere and, in Britain, remain common in Scotland and the north of England. Despite its hardy character and ability to withstand freezing winter blasts, the aspen has never had a particularly robust reputation. Its botanical name, *Populus tremula*, reflects its most striking physical characteristic. This is the trembling poplar, whose leaves are in constant motion. Even on an early September morning, when other trees are still as sleep, their heavy late-summer leaves all wrapped in soft mist, the aspen's long, slender stems will be shaking, finding a breeze that is not there. When John Keats was conjuring up an image of the ancient, overthrown forces of earth in his unfinished epic 'Hyperion', he described the leader of the defeated pagan deities as having faded eyes, a palsied tongue and a beard shaking 'with such aspen-malady'. The motion of the aspen suggests the aftermath of trauma, a quiet tremulousness, just strong enough to unsettle any sense of peace.

In some areas of Britain, the wood for Christ's cross was thought to have come from the trembling aspen poplar, and the trees have been shaking with guilt ever since. In his collection of oral poetry from the Scottish Highlands and Islands, the nineteenth-century folklorist Alexander Carmichael included an incantation beginning, 'Malison be on thee, O aspen tree! On thee was crucified the King of the Mountains!' Wordsworth's ballad 'Hart-Leap Well' tells the tale of a medieval hunt through Wensleydale in which the rapacious Sir Walter flogs his horses and dogs all day in pursuit of a deer, until his prey finally collapses and dies of exhaustion next to the spring where it was born. The tale of heartless human obsession leaves a permanent scar on the Yorkshire landscape, and centuries later, when every other trace of Sir Walter has vanished, the spot is marked by the 'lifeless stumps of aspen wood'. In fact, the aspen's capacity to rot makes it especially valuable to entomologists, because it provides a home for a host of rare insects, including the orange aspen hoverfly; but for those less concerned with the preservation of insects, these trembling, fungus-infected trees tend to inspire suspicion or pity.

The poplar has often been cast as a witness to brutal deeds. In classical tradition, when Phaeton plunged to earth after his disastrous failure to control the horses of the sun, his grief-stricken sisters turned into trees. Ovid describes the horrific process, as bark begins to surround their thighs, gradually spreading over their entire bodies until only their lips remain, 'vainly calling for their mother'. Although he does not specify the tree, the Italian setting led generations of artists to depict the metamorphosis of terrified young women into Lombardy poplars. Many centuries later, on the other side of the Atlantic, the poplar continued to feature in disturbing scenes. In Billie Holiday's well-known song, 'Strange Fruit', the 'Strange Fruit hanging in the poplar trees' refers to the black victims of lynching in the southern states of America, where the huge cottonwood poplar's strong branches would serve white gangs as a makeshift gallows.

Not all American poplars have such grim associations. The natural perfume of the balsam poplar, *Populus balsamifera*, is so heavenly that it became known as 'Balm of Gilead', the biblical aromatic remedy for all ills. The sun-loving Alamo cottonwood, *Populus fremontii*, which is found across the south-western states and Mexico, is a good source of vitamin C and has provided traditional medicines and treatment for wounds for the indigenous peoples of America. The common problem of misidentification by early settlers from Europe meant that there are also American trees going by the family name that are not really poplars at all. The beautiful, yellow-petalled tulip poplar, *Liriodendron tulipifera*, for example, is not a true relation, but because its craggy trunk and big, fluttering leaves were reminiscent of the poplars of the Old World, it was named accordingly.

The wonderful variety of poplars and their eagerness to interbreed has proved an irresistible challenge to modern plant scientists. In 2006, the black cottonwood, or California poplar, *Populus trichocarpa*, had the honour of becoming the first tree in the world to have its entire DNA mapped out. The international team of geneticists, based in California, hoped that full-scale analysis of the poplar's DNA would shed light on the gene structure of trees. The research is leading to practical experiments in tree breeding, with the multiple aims of combatting carbon emissions and plant diseases, and developing biofuels and biodegradable plastics. Genetic modification using lignin can lead to new kinds of poplar wood that break down easily, reducing the need for chemicals in paper production. Modified poplars also have a better capacity for phytoremediation, through which heavy metal pollutants are purged from industrialised areas. All these exciting possibilities bring their own risks, however, and the consequences of planting genetically modified trees are not yet known. Their effect on soil ecology or microbial communities has still to be fully tested and, given the poplar's reproductive energies, there are real concerns about whether GM trees could be kept under control if any unforeseen problems should arise. Feelings over this controversial issue were running so high in 2001 that the University of Washington's Center for Urban Horticulture was firebombed by activists, incensed at the idea of genetically engineered poplars.

LATE EIGHTEENTH-CENTURY MATCHSELLER

The poplar has always been the best tree in a blaze, of course, being a water lover. The high moisture content makes it burn slowly, so it is a good choice for heating ovens. In Shropshire, the floors and upper stories of the half-timbered Tudor houses were often built from local poplar wood, because it had a better chance of withstanding a fire. The flame-retardant qualities meant that poplar wood was quite the best for bellows, and for matches, too, being less quick to singe beards or fingers. A box of matches is so familiar, so domestic and yet so mass-produced that we hardly see these useful little mini-torches as remnants of a living tree. And yet, the young warriors of Ancient Rome wore wreaths of white poplar to show their allegiance to Hercules, who had crowned himself with the silver leaves of his own sacred tree, after triumphing over the fire-breathing, cave-dwelling monster, Cacus.

Poplar trees have many uses. The Lombardy poplar's capacity to provide a flourishing screen more quickly than most has long given it a useful role in landscaping, but with the proliferation of polytunnels, it can now hide all those shiny, plastic, Nissan huts, as well as protecting them from storm damage. The bushy boughs of these and other poplar species also provide shade for livestock and can be cut and fed to cattle. Recent research has discovered that hens feel safer in a shady field, so poplars, which do not offer easy places for hens to roost, provide a perfect environment for free-range flocks. Relaxed chickens are more inclined to relieve themselves in the fresh air, making for a healthier henhouse and even more vigorous poplars. Although the saplings grow quickly, established trees are often reinvigorated by the age-old practice of pollarding. This method of removing the tops of trees prolongs their lives as new shoots spring from the truncated stumps. A pollarded poplar can soon resemble a giant clipped poodle, with its bare bark sprouting round clumps of tight leaves. This may mean less shade, but it also means more cattle feed.

Poplars have always offered an easily renewable source of materials for joiners. The lightweight wood was right for shoe heels, clogs and wagon wheels, not to mention the bowls, trays and fruit punnets for which the wood is still in demand. The light colour of the timber also made it popular for floorboards. These trees provided living poles for vines and hops, while their twigs were made into brooms and the juice of their leaves was turned into remedies for earache. The poplar is still in demand: as the fastest growing hardwood in Britain, it is one of the few kinds of tree eligible for the UK government's Energy Crop Scheme, which encourages farmers to grow crops that will help reduce dependence on fossil fuels.

Despite the many benefits of commercial poplar planting, the timber trade has often caused distress to those whose relationships with trees are not primarily economic. When the eighteenth-century poet William Cowper saw that his favourite trees had been chopped down, he was horrified:

> The poplars are fell'd, farewell to the shade
> And the whispering sound of the cool colonnade,
> The winds play no longer, and sing in the leaves,
> Nor Ouse on his bosom their image receives.

Almost at once, the idea of the poplar became inseparable from lumber. Dismay over tree-felling is relatively common today, but this was probably its first powerful expression – powerful because it so obviously felt like a personal bereavement.

The peaceful atmosphere of Claude Monet's celebrated paintings of the tall poplars standing by the curving River Epte near Giverny is in many ways misleading, for he was prompted to buy the trees in the middle of his project when he learned that they were about to be felled. Monet had to bargain with the local landowners to allow the poplars a brief reprieve while they unwittingly fulfilled their roles as artist's models. As Monet arrested the movement of the river in his series of colourful impressions, so he secured for the doomed poplars a permanent place in posterity, where they continue to be admired, long after their original owners and the timber merchants have been forgotten.

Perhaps the most passionate response to the destruction of trees of this kind came from Gerard Manley Hopkins, who was working as a priest in St Aloysius Church in Oxford, when the poplars in the nearby water-meadows along the Thames at Binsey were cut down:

My aspens dear, whose airy cages quelled,
Quelled or quenched in leaves the leaping sun,
All felled, felled, are all felled;
    Of a fresh and following folded rank
        Not spared, not one . . .

Hopkins was deeply shocked by the destruction of an entire line of trees. For him, this was not just a matter of aesthetics, but an act of *spiritual* vandalism, because in his eyes, the distinct individual beauty of each tree was God-given. These trees were living expressions of divinity, singular and irreplaceable. In heavily industrialised Victorian Britain, Hopkins was acutely conscious of what he called man's 'smudge'; hacking down poplars symbolised the careless mutilation of nature by modernity. What would Hopkins have made, I wonder, of the engineered poplars now being designed to combat pollution?

HOLLY

I T is not easy to imagine a prehistoric earth – a planet entirely unshaped by human activity. From the merest petrified traces of extinct teeth and bones, sci-fi writers, film-makers and creators of computer animations have fed a collective fantasy world, where pterodactyls with outstretched claws scream through the thick forests, swerving past rope-like creepers, where shark-jawed creatures slide through strangely iridescent seas, threatening the inhospitable shores, where dinosaurs rage under the red shadows of dry, jagged cliffs. Here our imaginations run wild, secure in the knowledge that it is all reassuringly confined to a long since vanished and almost certainly fictional era. Yet what the fossil records also reveal is that among the strange plants flourishing in the Cretaceous Period, some hundred million years ago, was the holly tree. All at once, the remote age of the dinosaurs rushes closer. What place has this familiar tree in the world before time began? Palaeontologists have worked out that the Tyrannosaurus rex might glut its ferocious appetite on a passing triceratops, but many of its less ferocious contemporaries preferred the vegetarian option. And what could be more appetising to the thick-skinned, armour-plated, back-bladed spiky stegosauraus than the holly? And which tree could be better equipped to survive such attention?

The holly has been admired for its resilient qualities as long as there have been human beings with the capacity to record their views. The Ice Age survivor was celebrated by the Romans, who linked it to the old earth god Saturn, renowned for his command

HOLLY-WREATHED CONTENTS OF *A CHRISTMAS CAROL*

of the dark season. The ancient celebration of Saturnalia took place in December, around the winter solstice, when the days were short and the parties long. As other trees faded into mere skeletal forms, the holly, still covered in waxy green leaves which lock up water for the long, frozen, winter months, seemed ever more vigorous in comparison. Holly branches – bold, glossy and berry-bedecked – were brought indoors to brighten hearts and hearths. The holly tree is the Lord of Misrule, rebellious and ready for life, despite winter's worst. The legacy of this ebullient festival survived, of course, into Christian tradition, though Saturn's bold, pagan club and triumphal wreath were gradually intertwined with Christ's bloody crown of thorns and the promise of everlasting life. The plant's Anglo-Saxon name, *hollin*, fused easily with holiness and holidays to make it seem just the tree for celebrating Christmas.

As a consequence, the holly tree has the unenviable distinction of seeming unseasonal for eleven and a half months of the year. Though one of the shiniest evergreen trees, with its sleek, waxy, light-catching foliage, it remains largely invisible to most of us except during that fortnight in December, when suddenly it is everywhere. Undeterred by the pincushion leaves, we like to poke the scarlet berries around pictures and door-frames or to strew the emerald sprigs across windowsills. The holly's tenacity is evident as the gathered sprays retain their gloss even after days and days without water. If left undisturbed, a sprig of holly will eventually dry out, fading a little to a pale, olive green, but still losing none of its needle-sharpness. In profile, dried holly looks rather like a crocodile, just at the point when it is opening those deadly jaws. In spite of this, we determinedly welcome this tree into our homes each year and display the leaves on our front doors as a cheery greeting to visitors. The construction of a holly wreath does require a certain dogged perseverance, as the holly, furiously resistant to being forced into a circular frame, fights back with sharp defiance. Given the number of evergreen trees, it is odd that the prickliest of all has retained its sway over the festive season. But it *is* a very stubborn tree.

There is also something about those shiny, three-dimensional leaves, pointing up and down, in and out, that seems to set the mood. As soon as we begin decking the halls with boughs of holly, there can be no doubt about what will follow. In his seasonally astute comedy *As You Like It*, Shakespeare was quick to commandeer the most obvious festive rhyme – 'Hey-ho, the holly / This life is most jolly'. What people tend to forget, though, is that this apparently predictable chorus comes after the lines recalling the harshness of winter and the even more brutal effects of human failings:

Blow, blow thou winter wind,
Thou art not so unkind,
  As man's ingratitude.
Thy tooth is not so keen,
Because thou art not seen,
  Although thy breath be rude.
Hey-ho, sing hey-ho, unto the green holly.
Most friendship is feigning, most loving, mere folly.
  Then hey-ho, the holly;
  This life is most jolly.

Shakespeare's song of the jolly holly is not a cosy Christmas hit, but a call to arms against loneliness, despair and death. The holly tree reigns over the greenwood, offering refuge to victims of unkindness, betrayal and injustice, who discover, in turn, ways of combatting the situation. The holly's predominance in the dark season is not necessarily a temporary triumph of misrule, then, but a bracing rebuttal of the normal order of things. This is the tree that can prick the overinflated and provoke laughter in the face of the coldest blasts.

The holly provides winter fuel in the form of hot logs for human beings and high-energy food for sheep and cattle. The eye-catching berries sustain birds through the hardest winters. In areas where mistletoe is scarce, the mistle thrush is often known as the 'holm thrush' – or holly thrush – because it has such a soft spot for the fruit of this prickly tree. Under human influence, though, the holly tree has revealed a capacity for trickery and betrayal, because the bark, when stripped and soaked and boiled and fermented and pounded, turns into the stickiest substance. Birdlime, as this natural glue became known, can be spread over bare twigs in order to trap songbirds, which were – and still are – eaten with enthusiasm in many countries. It was not only thrushes, linnets and finches that were prone to sticking fast to birdlimed boughs: once farmers in the Lake District realised that their abundant stands of holly could furnish boatloads of the stuff, insects throughout the British Empire were under threat. Poacher's friend or natural pesticide? The sticky by-product of the holly tree was only as bad or good as those who applied themselves to profit by its various uses.

The wood of the holly tree has often been forced into disguise, too. Since exotic tropical imports such as ebony and mahogany could be prohibitively expensive, British woodworkers would often dye the hard, dense and very light-coloured wood of the holly tree. You might have a chess set fashioned entirely from holly, with dyed pieces pitched in battle against undyed compatriots. If you keep them in a Tunbridgeware box, it might also be made from holly, because the Victorians loved this fine, white wood. For daisy petals, butterfly markings, swallows' bellies or chequerboard patterns, for pale skies, light windows and snow-covered roofs, the wood of the holly tree was a crucial part of the marquetry palette, providing such a bright contrast to all those toning shades of gold, brick-red, dark chocolate and barley. Holly, as a hardy native, was relatively easy to find. Resistant enough to stand in for boxwood, engravers who were struggling to make ends meet could rely on woodblocks cut from a holly tree. The handles of their tools and teapots were often derived from holly, too, while lace-makers drawing fine light threads into intricate designs used bobbins made from this pretty white wood.

The tree itself, though so widely recognised, has been subject to mistaken identity. Cawdor Castle in Inverness-shire is famous not merely on account of Macbeth, but also because of the legendary thorn tree, which grew for years inside the ancient stronghold. According to local legend, an early thane of Cawdor had a dream in which he was told to load a donkey with gold, send it out on its travels and follow to see where it stopped, because that was where he should build his castle. So the thane did as the dream decreed and when the donkey stopped to graze by a thorn tree, the site of Cawdor Castle was fixed. Centuries later, when Dr Samuel Johnson and his young travelling companion, James Boswell, visited Cawdor during their journey to the Hebrides, they were astonished to see the ancient thorn rising 'like a wooden pillar through the rooms of the castle'. They would probably have been even more surprised to know that in the twenty-first century, DNA testing would reveal that the legendary thorn of Cawdor was in fact a holly tree.

Although the discovery may have unsettled the old legend and songs about the Thorn of Cawdor, the fact that it was really a holly tree all along may not have been such a bad thing – especially at the legendary home of Macbeth. For among the holly's traditional roles is that of sturdy defender against witches. Holly berries were often planted outside homes to ensure that a strong, slow-growing tree would be there in perpetuity, to ward off malevolent visitors and evil spirits. Hollies have always been popular for hedging, partly for practical reasons, since their tough, prickly natural growth tends to deter intruders, and partly as a result of a deep-rooted belief in their protective powers. Richard Mabey's magisterial *Flora Britannica* records some of the lingering beliefs surrounding these trees. In Buckinghamshire, there was widespread resistance to cutting down holly trees, for fear of unleashing witches, while in East Sussex farmers leave intermittent holly trees to grow tall above the rest of the hedge in line with the old local belief that these trees prevent witches from running along the top. Even when hedges are grubbed up to extend the size of fields, holly trees are often left alone, perhaps as a record of the old boundaries, perhaps to avoid any mysterious, sinister consequences.

How can you tell a true holly from a false one, when almost every tree with spiky leaves has been referred to as a 'holly' at some time or another? The holm oak or holly oak is no real relation of the holly tree, but still takes its popular and botanical names (*Quercus ilex*) from the holly (*Ilex aquifolium*), because of its evergreen character and needle-tipped leaves. The kermes oak, *Quercus coccifera*, too, with its tough evergreen leaves and resident scarlet beetles, has also been confused with the holly tree. Neither sea holly nor knee holly is any relation of the holly tree. They both belong to completely different species: one is a kind of thistle, the other a kind of broom. As soon as a tree becomes identified by a particularly striking

HETEROMELES ARBUTIFLORA, ALSO KNOWN AS THE TOYON TREE

feature, such as a blood-red berry or a spiky leaf, any other plant bearing something similar is likely to be named after it, irrespective of what the botanical experts may decide.

When European pioneers finally reached the west coast of America, for example, they found an indigenous tree with bunches of scarlet berries and evergreen leaves that reminded them of similar species back home, and they christened it California holly accordingly. The species in question is now classified as *Heteromeles arbutiflora* and is popularly known as the toyon tree. Perhaps it was all for the best: the American film industry took its lead from the original mistake. 'Frankie goes to Toyon wood' would not have had quite the same ring. 'Hollywood' is the perfect name for the glossy world of Californian film-making, where tinsel does not have to be confined to two weeks in December. Even those great iconic hillside letters are oddly reminiscent of the white timber of the holly tree.

The long history of holly confusion is not at all surprising because, in addition to all these holly wannabes, there is a huge range of different varieties. In some kinds of true holly, the leaves are variegated rather than plain, and the distinctive green looks like oil paint quickly brushed across a creamy base. Sometimes, a golden or silver outline can just be discerned, peeping out like a circus tent on the far side of summer trees. Some of the trees have smoother leaves, especially in maturity, when the upper branches become less spiky, leading people to regard the holly as an emblem of the mellowing that comes with age. There are types of holly that grow into towering, conical mountains of green, with leaves so smooth that they are often mistaken for exotic, imported evergreens rather than hardy natives. Others, though, such as *Ilex aquifolium* 'Ferox', or the 'hedgehog holly', have spines all over the surface as well as round the edges of the leaves. Not all holly berries are scarlet, either – there are kinds that cluster in yellow or gold, such as 'Bacciflava' and 'Golden Girl'. The fruit of the inkberry holly, on the other hand, is, as the name suggests, black.

The astonishing variety of holly trees has inspired lovely popular names, like the 'Golden Milkmaid', with the waviest leaves, or the 'Moonlight Holly', with its cloud of pale light, or the rather less lyrical 'Leather-leaf'. Many have regal associations, no doubt in tribute to the holly's gemlike berries and precious-metal crested leaves. The popular names can be misleading, though: 'Golden Queen' and 'Silver Queen' are both male varieties while 'Golden King' and 'Silver Milkboy' are female. In botanical terms, the holly is diœcious, which means that male and female flowers grow on separate trees. This is why we are usually advised to plant several hollies in order to ensure a good crop of berries, because the fruit only grows on the female tree, after she has been fertilised by a nearby male. Mature holly trees are resistant to transplanting, so it is always important to plant a group of young saplings with plenty of space for growth, to avoid inadvertently leaving isolated widows as the trees spread to attain their full height and extent. In their natural habitat, older trees will keep their distance and so, at the Stiperstones in the Shropshire hills, the ancient hollins (or holly woodland) known as 'The Hollies' spreads out across the bare ridge to allow each gnarled veteran room of its own.

Although the holly's most striking feature, its bright berry, belongs to the female of the species, this tree has always been a symbol of masculinity. The rebellious, spiky, holly king was the traditional winter rival of the stately oak king, who ruled over the summer months. In the medieval Christmas tale of *Sir Gawain and the Green Knight*, Sir Gawain's formidable green opponent carries a holly bough charged with the power of this ancient lore. As we know from the familiar Christmas carol, of all the trees that are in the wood, it is the holly that bears the crown. The winter king only rules until Twelfth Night, which, in many areas of Britain, was marked by spectacular bonfires built from redundant holly decorations. Anyone who has put a match to the festive leaves in the New Year knows just how combustible they are, and so it is not hard to imagine how dramatic the annual torch-lit procession must have been at Brough in Westmorland, where on 6 January a whole holly tree was ceremoniously set alight and paraded through the black night.

Nevertheless, in many areas, the later winter festival of Shrove Tuesday was celebrated by a carnival of 'Holly Boys', or figures crafted from holly, which were conducted through the village streets. During the parade of the 'Holly Boys', the 'Ivy Girls' were

TORCH-LIT PROCESSION AT BROUGH IN WESTMORLAND

supposed to keep well away. A male holly tree can, after all, fertilise any number of nearby females and not everyone wants a crop of berries. It may well have been this potent reputation that drew Robert Burns to the holly tree. In his vision of a very welcome visitation by the Muse, the dazzling beauty comes to his cottage with 'Green, slender, leaf-clad *Holly-boughs* . . . twisted, gracefu' round her brows', and, after setting out his duties as a Scottish bard, she concludes with the coronation of her devotee:

> *And wear thou this*, – She solemn said,
> And bound the *Holly* round my head;
> The polish'd leaves, and berries red,
>   Did rustling play;
> And, like a passing thought, she fled,
>   In light away.

For Burns, the vibrant, virile, native holly was wholly superior to classical laurels for his accession to the role of Caledonia's bard, especially as its mischievous, self-mocking character was guaranteed to puncture any pomposity. More than fifty years after Burns' youthful vision, the elderly poet laureate, William Wordsworth, by then in his seventies, took a handful of berries and planted them at Mrs Fletcher's home in Easedale, on the edge of Grasmere, in memory of his holly-crowned hero. Visitors to the Lakes can still comb the grounds of Lancrigg for berries from these very literary trees.

Today, however, the holly's reputation has been balanced by much more feminine associations. It is now a popular girl's name, especially for little girls born in late December. Holly Combs and Holly Madison were both December babies (though a number of other celebrity Hollys seem to have been born in the spring). The name 'Holly' became chic in the 1960s, after Audrey Hepburn captivated the world in *Breakfast at Tiffany's*. In Truman Capote's original draft of the story, the central character was called Connie Gustafson but, luckily, he changed his mind and she became

AUDREY HEPBURN AS HOLLY GOLIGHTLY

Holiday Golightly, better known as Holly. Anything less like the stubborn, slow-growing, prickly evergreen tree that hates to be transplanted than Capote's flighty heroine is hard to imagine. As inspiration for a colourful, inviting, counter-conventional and very decorative free spirit, who is not quite all that meets the eye, however, the holly tree may be somewhere in the background.

SYCAMORE

Many rail companies now run special 'leaf-fall timetables' from early October, routinely warning passengers that trains may arrive at their destinations 'up to three minutes later than shown'. It is a response to a problem that hit the headlines in the 1990s and continues to provide a seasonal story to intersperse with the party conferences – 'leaves on the line' or, as it is known in America, 'slippery rail'. As showers of moist leaves land on railway lines every autumn and become compressed, the slick-coated tracks can cause trains to skid. The clearance operation takes time and costs money, but it is essential work, because the trains cannot run safely until it has taken place. For commuters already faced with a tedious, expensive, daily trail, the idea that the journey home might be even longer is not at all welcome, and so in Britain 'leaves on the line' has become a catchphrase for the inability to cope efficiently with the vagaries of the weather – as well as something of a standing joke. It has even furnished the title of a self-help guide, *Leaves on the Line: How to Complain Effectively*, as well as an anthology of letters sent by disgruntled passengers to the *Daily Telegraph*. But where do these annoying leaves come from? Mostly, it seems, from the sycamore tree.

These familiar, bushy-boughed trees line many suburban railways, screening passengers from unsightly derelict warehouses, and urban gardens from the rattle of passing trains. But as the season changes, they are exposed as turncoats. Gone are the lush, green friendly attendants waving us on, and in their place stand gaunt skeletons, mocking our thwarted progress. When William Blake observed that 'Not everybody sees alike: a Tree that moves some to Tears of Joy to other Eyes is just a green thing in the way', he might well have had the sycamore in mind – for there is probably no other tree that divides opinion so dramatically.

If the soft, flat, open-handed leaves of the sycamore can cause an autumnal rush of commuter blood pressure, its oozing globules of sap seem to provoke even stronger feelings. The desperation of gardeners is only too plain from the Royal Horticultural Society website, where one anguished comment reads 'the SYCAMORE – aaaaargh! It rains leaves and now sap and insects of all kinds emerge from it. The sap is everywhere and makes all the garden furniture sticky . . . it's just a smelly, sticky nightmare!' This contribution triggered a string of suggestions for how to *deal* with the sycamore – which clearly looms large in many minds as a Problem Tree. Nor is this a peculiarly modern irritation: in 1664, when Evelyn published *Sylva*, the sycamore was already causing concern: 'the Honey-dew leaves . . . fall early and turn to a mucilage'. This putre-fied, brown mush was not only messy and unattractive, but also home to 'noxious insects'. Sycamore leaves are covered in 'honey-dew' because this sticky substance is excreted by the swarming aphids that live on the abundant sap. Though increasingly planted for the purpose of creating attractive, shaded paths for promenading, in Evelyn's eyes sycamores did nothing but 'contaminate and marr our walks' and should, accordingly, be 'banish'd from all . . . Gardens and Avenues'. This is an early version of both the sticky sap night-mare *and* 'leaves on the line'.

And yet, the very idea of 'honey-dew' leaves has an almost magical, milk-of-paradise quality about it. For the poet John Clare the 'splendid sycamore' with its mountain of massy, sunny green foliage was one of spring's adorning glories. No matter that it had no place in polite company, for Clare it was an aristocrat of the fields, a 'rich beauty' ready to share its opulence with all comers. The sweet sap and sticky leaves were not to be regarded as a nuisance to gardeners, but as a great gift to the world, and he bid people listen to the humming insects and the 'merry bees, that feed with eager wing, / On the broad leaves, glaz'd o'er with honey-dew'. You can almost hear the bees homing in on those controversial leaves – and the poet, who aligns himself with these 'happy Ariels', sucking inspiration and fuelling his own imaginative flight from the delicious resources of the immediate, natural world.

Though often given to nostalgia, Clare was well ahead of his time in his understanding of the sycamore's contribution to the ecosystem, though it has taken a while for people to realise the true importance of those apparently commonplace bees as pollinators and therefore food security guards. As an aphid megalopolis, the sycamore also nourishes masses of swallows and house martins, not to mention blue tits, garden warblers, robins and chiff-chaffs. It is both an apian and an avian cafeteria. This may all seem small consolation to someone whose garden has been turned into a bug-house or bird latrine by the presence of an oozing sycamore, but the tree plays a leading part in the natural economy of the garden. Sycamore sap can also be extracted and then boiled to make syrup or left to ferment and turn into wine. Those with a self-sufficient or frugal impulse can mimic the bees rather than their sap-averse neighbours by tapping into these goo-filled trunks.

For many, the sycamore is too generous a tree altogether. It disturbs people's sense of proportion and even seems to uncover lurking puritanical anxieties about excess. It is the tree of profusion: too much sap, too many leaves. Too many sycamores, in fact. At every turn, the rude health and vigour of this tree count against it. Sycamore wood is as hard as oak, but far less highly prized. It is a good choice for rolling pins and wooden spoons, because its light, fine grain looks smooth and bright and fresh in the kitchen, as well as being easy to clean, so for affordable kitchen items, it cannot be beaten. Rolling pins do not command the respect afforded a tall, oak-built ship or a finely carved mahogany sculpture, however. This is the wood of the kitchen, not the grand dining room, the chopping board rather than the board room. For some years, it had a claim to fame in Wales because the largest love-spoon in the world, over twenty feet in length, was carved from a single sycamore branch. This magnificent spoon created by Ed Harrison in 1989 is still on display in Cardiff, though it lost its title in 2008 to his spectacular red cedar wood love-spoon, which is twice the length and now reclines in splendour at Caerleon.

As sycamore wood is never in short supply, it lacks any rarity value. In fact, a powerful, multi-faceted objection to the sycamore is that it is rather common. This vigorous tree springs up everywhere, spreading rapidly because of its distinctive propeller-like seeds. The little transparent boomerangs seem eager to fly off the twigs, gliding with every gust and eddy, racing to see which can fly furthest. The seedlings sprout easily, and don't seem to mind much whether they have found a home on the edge of a well-groomed lawn or right in the middle of a lovingly tended rose bed. These perky little shoots are usually treated as weeds, and yanked out of the soil before they can get their roots down.

The sycamore is, indeed, one of the few *trees* to feature in Richard Mabey's brilliant book *Weeds*, which identifies another of its apparently off-putting features: 'It is a foreigner.' Unlike those respectable oaks and ashes, elms and yews, the sycamore is widely regarded as a non-native species, introduced into Britain in the late fifteenth century. And this might not matter much to anyone, were it not for the sycamore's astonishing ability to multiply. Rapid in growth, rapid in expansion, the sycamore strikes many as an enemy invader, driving out our poor native trees, depriving our wild flowers of light with its thick canopy of flourishing leaves. It is the grey squirrel of the vegetable world, no sooner here than it was everywhere.

Except that there is a possibility that it is a native after all. In Christ Church Cathedral, in Oxford, the thirteenth-century shrine of St Frideswide has carvings of sycamore leaves, which has led to further debate over the sycamore's alien status. In the legend, Frideswide hid among the trees to preserve her vow of chastity and escape the pressing pursuit of the king of Mercia. Whether the medieval sculptor had local species in mind when portraying her, peeping from a circle of leaves, on the carved boss above her tomb is now rather hard to judge. The five-lobed, palmate leaves on the cornerstone may have been chosen for symbolic reasons, to evoke the stigmata, or Five Holy Wounds, of the crucifixion, especially as they resemble open hands. Archaeological evidence is also fuelling the debate over the sycamore's origins, since its pollen, when fossilised, is difficult to distinguish from that of the native field maple, clearly part of the fauna of these islands for thousands of years.

Even if sycamores have only been growing in Britain since Tudor times, they are hardly recent arrivals – descent through five centuries would provide a respectable enough pedigree for most of the people who mind about such things. (The ancient sycamore at Scone Palace was supposedly planted by Mary Queen of Scots, which might have bestowed some sense of nobility.) And if the sycamore is really such an aggressive expansionist, how have all those other trees managed to survive here for so long? There has been no marked increase in sycamores in Britain since the early 1970s and, as the seedlings thrive in lighter situations, the dense foliage of the mature tree means that sycamores are self-regulating, growing more readily in the gaps between other trees than in solid woods of their own. Matters of origin are, in any case, subject to fashion. In the nineteenth century it was the exotic imports that were in demand, as the horticulturalist John Loudon commented: 'No residence in the modern style can have a claim to be considered as laid out in good taste in which all the trees and shrubs employed are not either foreign ones, or improved varieties of indigenous ones.' For fashionable Victorians, the question of which trees were natives had very different implications. The English garden was becoming a symbol of national internationalism: as the empire expanded, so did the variety of trees, and the more far-flung their origins, the better. But by then the humble sycamore was too familiar to make a comeback as an exciting stranger.

SHRINE OF ST FRIDESWIDE, CHRIST CHURCH CATHEDRAL, OXFORD

Home-grown or imported, the sycamore's adaptable nature means that it will always be quick to rush in where other trees fear to tread. Sycamores grow quickly and reliably, taking the hard edges off new roads and housing estates, cushioning children's playgrounds with greenery. You can see rows of them on exposed Cumbrian hillsides, beside the M6 and the West Coast line, linking arms like Beryl Cook partygoers, all decked out in emerald frills and ready for a good time, whatever the weather. These hardy trees are well equipped to survive the exhaust fumes of the inner city and will grow almost anywhere, though heavy clay is not the most congenial soil.

They withstand the bracing salt-saturated air of northern coastlines better than any other broadleaf. In the North Yorkshire spa town of Scarborough, the cliffs bearing the brunt of the winds coming off the North Sea are screened by tall, hardy sycamores, while in Edinburgh these trees reach enormous heights above Waverley Gardens. On small Scottish islands, not far from land, sycamores line the curving shores, striping the view of the distant hills in graceful tiers: from the stout, scabbed veterans in the damp woods to the lithe, young trees that seem poised to step out into the freezing water of a sea loch. The movement of trees in the rather frequent breezes is mirrored in the rippling tides, while those big palmate leaves, surging together on the banks, wave back and forth, and back again.

Sycamores have spread from Europe to many parts of the globe, from Chile to Tasmania, from Canada to the Canary Islands. The forests of Southern Australia are home to huge stands of sycamore, which have taken very well to the warm sun and soaking rain. In New Zealand, they seem to have retained some of their Old World reputation, as they grow anarchically in wasteland and along the sides of roads. The tree known as the sycamore in the United States is the *Platanus occidentalis*, which is every bit as vigorous and grows even taller than its European namesake. In the remote wilderness of Sycamore Canyon in Arizona, a spectacular waterfall cascades over semicircular tiers of red rock, passing through thick, waving crowds of sycamore trees.

Undaunted by any environment, it seems, the sycamore is one of the world's most travelled trees. When *Apollo 14* blasted off in 1971, one of the astronauts, Stuart Roosa, took some seeds with him on the mission to see how they would be affected by going outside the earth's atmosphere. After orbiting the moon thirty-four times, the seeds landed back on earth and were monitored by NASA scientists. Forty years later, moon sycamores are still flourishing across the USA, as living testaments to the courage and ingenuity of human-kind and to the remarkable resilience of this supposedly 'common' tree.

There is something inspirational about the apparent ordinari-ness of the sycamore. In Ireland, the famous wishing tree of Mountrath was an old sycamore, whose cracked, crazy-paving of a trunk proved irresistible to visitors nursing secret hopes and desires. For years people came to make their wish, driving it home with coins and nails until the tree shone like an elderly silver-scaled dragon, too weighed down with everyone's cares to be able to get airborne again. Unfortunately, all that human expectation eventu-ally killed the tree. In one of Heaney's moving elegies for his mother, he imagines her as a dead wishing tree; but instead of conveying despair, the old tree unexpectedly sails off to heaven, sloughing off all the nails and coins in a marvellous vision of uplift and consola-tion. This is very much a tree capable of eliciting tears of joy rather than just being 'a green thing in the way'.

When Percy Bysshe Shelley and his wife, Mary, were living in Italy in 1819, they witnessed the fall of sycamore leaves in the woods around Florence. It was a moment of profound personal misery, prompted by the death of two of their children, William and Clara, further deepened by the public dismay over the infamous Peterloo Massacre, when the British government had so brutally suppressed a peaceful, open-air, political meeting in Manchester. Shelley, still little read and far from English audiences, felt his own leaves to be falling as rapidly as those of the 'oozy woods' around, which were now suddenly turning 'grey with fear'. In spite of this sense of all-consuming dismay, he held on to the hope that the sycamores' withered leaves could 'quicken a new birth'. The west wind that inspired his great ode was not only driving dead leaves to their final destina-tion, but also propelling their 'winged seeds'. However bleak the immediate prospect, however brown the surrounding leaf-mould, it could only be a matter of months before the dreaming earth would wake once more and so the poems ends memorably: 'If Winter comes, can spring be far behind?' Like so many before and since, Shelley found comfort in the sycamore, that unlikely, commonplace companion, whose annoying oozings and messy mucilage were really holding the promise of something better to come.

Though Shelley, as a self-proclaimed atheist, would not have welcomed the suggestion of Christian influences on his thinking, the sycamore was well known for its biblical associations. This was the very variety of tree thought to have been climbed by the unpopular tax collector Zacchaeus, when he wanted to see Jesus without being spotted by the crowd who were obscuring his view. In fact, the biblical sycamore was a kind of fig tree (*Ficus sycomorus*), which still grows freely in Middle Eastern countries. In its journey through languages, the *Ficus sycomorus* became the 'sycamore' in English, and so the literary character of the English sycamore, *Acer pseudoplatanus*, was coloured by the older, biblical meaning. When Wordsworth describes resting under a sycamore on the banks of the Wye above Tintern Abbey, or when Izaak Walton selects the shade of a riverside sycamore as a spot for contemplation in *The Compleat Angler*, their choices may suggest the possibility of a glimpse of the divine.

Whether or not Wordsworth or Walton had Zacchaeus in mind, they both described the sycamore as 'dark'. This description may reflect the alternative etymological tradition, which recognised the *Ficus sycomorus* as a kind of fig tree and thereby encouraged links between sycamore leaves and the Fall of Man. It seems rather more likely, though, that both authors were really thinking about shade. For one of the most enduring and well-attested aspects of the sycamore is that sun-blocker foliage. Those great clumps of grass-green leaves, sprouting like huge heads of broccoli on their tall, elegantly spreading trunks, cast motley shadows over sunlit lawns. Whatever Evelyn thought about the sycamore's less appealing habits, his very objection reflects the tree's reputation as a provider of cool, shady walks. And perhaps this helps to explain the sycamore's frequent appearance in a world of wishes or private fantasies. For how often do people really *need* heavy shade during the average British summer? The mere mention of a 'dark sycamore' conjures up a gloriously hot July day – the kind that shines in everybody's dream of summer, but so often fails to materialise for the barbecue or school fête.

Sometimes summer sunshine is real enough, and the shade of a full-leafed sycamore has always offered more practical help to those routinely exposed to the elements. Farm workers faced with the strenuous tasks of clipping sheep or picking fruit and those cutting hay or harvesting crops, without the protection of an air-conditioned tractor cab, turn to the dark sycamore as a great natural parasol. For anyone involved in construction work or summer sports, for all those serving on seasonal stands at summer events, markets and trade fairs, a handy sycamore is a haven from the relentless rays and, as such, its commonplace character is only too welcome.

In the small Dorset village of Tolpuddle, the old sycamore on the village green still stands as a memorial to the agricultural labourers who gathered under its shady branches in 1834. They were in need of a drop in temperature and a rise in pay, after suffering wage cuts which meant they were no longer able to support their families. When a few of the men agreed to band together and demand fair pay, they were charged with swearing an illegal oath, tried, found guilty and transported to Australia. Public outrage over the case of the Tolpuddle Martyrs eventually resulted in an official pardon, and at last they were allowed to return home. The episode came to be seen as the beginning of the Trade Union Movement and the fat old Dorset sycamore was gradually turned into a site of political pilgrimage. The Tolpuddle sycamore is another symbol of hope for those whose lives are far from easy; but, as so often with this oddly divisive tree, this association has been more welcome in some quarters than in others.

THE TOLPUDDLE MARTYRS

Despite the hard reality of agricultural labour, the image of a flourishing tree on the village green is also part of an old, pastoral ideal of eternal summer, where work, sweat and poverty have no place. And this gives us fresh insight into the annual aggravation of those leaves on the line. For underlying the widely felt annoyance of being late for work is some lingering dream of endless summer holidays, all innocent of alarm clocks and electronic diaries. 'Leaves on the line' is an adult equivalent of 'back to school' – that ominous signal of returning routines, demands and uniformity. It is a reminder of shortening days and diminishing temperatures, of disappearing sandals and things closing down. No wonder a delayed train seems the final straw.

BIRCH

M Y grandparents' garden was a paradise of bright flowers and seemingly endless sunshine. What made it all the more magical was being hidden away behind a high wall, so that anyone walking past along the street had no inkling of how close they were to another world. In the autumn, the graceful upper branches of the mature birch tree standing guard over the perimeter wall did scatter hints onto the wet pavement, but they turned almost instantly from gold to brown and then disappeared. Safely cocooned within heavenly scented shrubs and warm, herbaceous borders, we were oblivious to the rhythms of the working day and would play hide and seek, make castles from empty cream cartons, or take turns on the rocking horse, until we had had enough of the fresh air and each other. I was not very old, though, when I scrambled onto the big pile of logs by the wall and up into the birch tree. Just as I was climbing high enough to see out into the street, my foot slipped on a smooth, damp branch and, catching at the top of the wall to stop myself from falling, I saw the black, spiked fist of barbed wire which ran along the top an instant before it scored my arm. For the rest of the day, sore, shaken and ashamed, I took dubious comfort from showing off the biggest plaster in the box. The cut healed, of course, but it left a long red line running down from my wrist, which very gradually faded into a smooth, white thread of a scar. It is still there, lightening in the summer as the surrounding skin goes brown. The little accident left a permanent portrait of the birch – a white, gently curving reminder of unheeded warnings and unexpected admonishment.

BIRCHING

For such a seemingly soft and inviting tree, the birch does command a fearsome reputation. Its name is synonymous with corporal punishment, because the flexible twigs inflict the sharpest reprimands. Although 'birching' was regarded as a more benign form of punishment than the terrible 'cat o' nine tails' it replaced, those serving on British naval vessels in the nineteenth century still lived in fear of being flogged with birch rods. Such brutal beatings continued to be meted out according to judicial sentences in English courts until 1948. Schoolboys guilty of serious misdemeanours were also subjected to the painful and humiliating experience of having their buttocks birched – apparently this was thought to be character-building. But, I wonder, what sort of character? The lingering folk belief that saw birch twigs as a means to whip out evil spirits may have impelled the strong arms of schoolteachers, or at least furnished some self-justifying excuse for their sadistic inclinations. The idea of beating children into knowledge runs so completely counter to current attitudes that it is hard to believe that this savage maintenance of authority was widely accepted as legitimate. And yet biographies, memoirs and autobiographical novels are full of excruciating memories of school floggings and unjust punishments, routinely meted out. In William Shenstone's once surprisingly popular poem 'The School-mistress', the pupils learn their lessons under the shadow of a vigorous birch tree that grows in the playground, endlessly sprouting fresh instruments of terror. It is easy to imagine how the thin, wispy, black boughs of a birch tree in winter, flying about in the coldest blasts, might haunt the dreams of a frightened child.

The power of birch rods is also part of the rather mixed legacy of Ancient Rome. In classical days, lictors demonstrated judicial authority by wielding a ceremonial axe braced in a formidable bunch of birch rods. The fasces – or bundle – has been seized by political parties ever since. For the French revolutionaries, this icon of the Roman Republic not only represented strength in unity and freedom from hereditary rule, but also, as a bundle of twigs, something well within the grasp of ordinary people. The birch rods meant people power – until seized by their rather more ambitious representatives. Later, Benito Mussolini derived both personal strength and a political party from the fasces, which featured in a highly stylised design on his black flag. There was nothing flexible about these tightly bound rods or the ominously protruding axe, especially when laid horizontally as a strong perch of parallel lines for the spreading symmetry of the Fascist eagle. In the United States, the classical emblem of the axe and birch rods has continued to symbolise judicial authority. It is prominent in the insignia of the National Guard, as well as in the frieze above the Supreme Court Building. Above the door into the Oval Office, too, the fasces are a reminder of the enforcement of justice to all who enter there.

Birch rods have sustained a formidable role in public and private life, though the bundles of twigs traditionally found in well-run households are not generally axe-scabbards. Even when domesticated into brooms, however, they retain some of their intimidating qualities. Birch besoms whisk away the cobwebs and the cat's hair. They even banish the dead leaves to which they might once have been attached. Whether serving as a midnight steed for witches or as a weapon to ward off donkeys from the lawn, the birch broom is often an instrument of power. It can be a force for good, of course, banishing unwanted detritus and clearing clutter to enable a fresh start. The New Broom, beloved of political cartoonists, sweeps into dusty old corners and sends the traditionalists scurrying in all directions.

The elegant trunks of birches, too, so slender and fair, are agents of surveillance. Take a careful look at those dark markings and you will see how they follow you about, like tired eyes, hooded and hollowed. This Argus tree is watching, so be on your guard. And the bark peels off like flakes of skin – that pale, cream, pitted surface is all smooth on the other side, flesh-coloured, and pocked with little slit-shaped spots. Stripping a birch is like slowly separating the pages of an old, damp book and then being unable to make any sense of the characters.

In mixed woodlands, the common birch, overshadowed by taller, fatter, broadleaves, is not at first particularly inclined to attract attention. This is another feature that makes these trees so unsettling: birches are the ugly ducklings of the woods, for the unassuming, gangly, brown saplings gradually change into graceful swan-necked trees, their branches arching into shimmering leaves. Admittedly, as they age, they begin to lose some of their sparkle – a clump of elderly silver birches can resemble a disconsolate zebra, tall, forlorn and ankle-deep in mud. Not all birches do turn silver, however. Another species common in Britain is the browner, fluffier, downy birch, which, according to its botanical name, *Betula pubescens*, seems doomed to remain in the difficult teenage years for its entire lifespan. Perhaps that is why the downy birch is forever reaching upwards, as if straining to appear taller than it really is, unlike its more graceful, gently cascading relation.

Betula pendula has not always been known as the silver birch. At one time it more often went by the names of 'white birch', 'lady's birch' or 'weeping birch'. 'Silver' seems to have taken root only in the nineteenth century, from seeds scattered in poems and popular songs. It was the Canadian poet Pauline Johnson who really fixed the colour of the birch in the twentieth century, when her lyrical recollections of the native forests of her Mohawk ancestors turned into a staple of primary schools and scout camps. Ever since, Canada has been 'land of the silver birch, home of the beaver'.

Birches are one of the last trees to shed their annual foliage, sometimes still clinging onto a relic of the previous summer as the butter-coloured catkins appear to greet the spring. In November 1805, the month of Trafalgar, Dorothy Wordsworth was exploring the woodlands near Ullswater with her brother William and was especially delighted by the 'lemon coloured' birch. Her description may make us wonder about the age of the lemons she was used to, because they would have had to be very ripe to match the deep colour donned by the birch tree for its final seasonal fling. We call these birches 'silver', but they are often more like polished bone, studded with pewter and, with the Midas touch of autumn, showered in gold leaf. On a bright day in October, a silver birch can light up a hillside like a waterfall of frozen gold.

It is not surprising that birch trees have inspired artists and designers. Their pale, shapely forms provide a perfect contrast to almost every other natural shade, so nineteenth-century artists loved to make them cut across their subtly coloured canvases. John Ruskin caught a silver birch perching on smooth rocks by a mountain stream, while his sometime friend, John Millais, capitalised on its shining bark to offset both the glistening skin of the naked woman who is tied to its trunk and the polished armour of the knight who frees her with his sword, in the moonlit painting *The Knight Errant*. This was the only nude Millais ever exhibited and it caused uproar in the Victorian art world, because his almost photographic treatment of the contrasting textures rendered the entire scene unacceptably realistic.

At Sundborn in Sweden, the home of Carl Larsson is still flanked by silver birches, which visitors recognise at once from his

JOHN EVERETT MILLAIS, *THE KNIGHT ERRANT*

paintings, in which the natural waves of the native trees make perfect settings for his delicate watercolours of family life. Wisps of fair hair lift to touch the lower twigs, as they each blow in the summer breeze. Larsson's contemporary, Gustav Klimt, also celebrated the dazzling white and golden canopy of the autumnal birches in Austria, in accentuating the trunks to render *Birkenwald* (or 'Birch-wood') an almost abstract design of parallel lines and glowing colour. Throughout the 1930s, the birch tree found its way onto plates and pots, earrings and endpapers, curtains, cushions and shawls, its elegant curves and distinctive monochrome making it an emblem of Art Deco. There were even small, ornamental birch trees for windowsills, fashioned from polished metals and copper wire, with dozens of tiny mother-of-pearl petals to catch the sunlight.

The beauty of birch trees is best appreciated, however, in their natural habitat – poised on a northern hillside, clustering along the edge of a damp moor. These are trees to admire as stand-alone specimens, sleek as herons by the side of a twisting burn, or to celebrate in woodlands as they gather shyly, staying far enough apart to let the light shine through. In the Himalayas, where everything is magnified and elongated, they look like cracks in the vast mountainsides, opening for a moment to let the breath escape. When you get closer, the trunks of white birches stretch like long ropes let down from heaven, but the branches seem to whisper a warning – that if you were to climb up into the clouds, you might never come back to tell the tale.

In northern folk tradition, birch trees, or birks, are borderers between the known and unknown worlds. The graceful, sinuous birks, their lissom trunks bending almost into question marks, may seem to be issuing invitations to trysting ('Bony lassie will ye go to the birks of Aberfeldey?'), but all too often the stories associated with them take unexpected turns. The old ballad of 'The Wife of Usher's Well' tells of a mother who sends her three sons away to sea, only to spend the rest of her days wishing for their safe return. When they do come home at last in the depths of winter, 'their hats were o the birk'. Whether this means that their hats are trimmed with birch leaves or made of birch bark is beside the point, for the ballad goes on to tell us that the birk was not of earthly growth, but thrives 'at the gates of Paradise'. The lads have come back from the dead to visit their bereft mother, and though the birch may be a sign that they have gone to a better world, in some versions of the ballad their fate is more mysterious altogether.

Birches certainly cut ghostly figures on a moonlit night, haunting the woods and processing down hillsides. For Sorley Maclean, the empty birch-woods surrounding a deserted settlement on the tiny isle of Raasay, off the coast of the Isle of Skye, were ghosts of the girls who once lived there, silent, straight-backed, with bent heads. In his haunting poem 'Hallaig', he imagined them following the old paths from the land of the living to fill the steep slopes, 'their laughter a mist' in the ears, 'their beauty a film on' the heart. The trees of Raasay, like many birch-woods, beckon to a lost world, beyond the mist, to a clearing on the far side.

There are times when people want to escape from the world for a while and, as Robert Frost put it in his well-known poem 'Birches', 'then come back to it and begin over'. His imagined route to an alternative place was via the trunk of a birch, climbing 'black branches up a snow-white trunk / *Toward* heaven'. It was only a temporary ascent into enlightenment, though, reassuring because the upper branches of a birch were so thin that they would eventually dip to set him down safely on the earth once more. The birch's admonishment need not be harsh or frightening: it can take the form of a gentle reminder of what is right about the world we know.

The mysterious birch tree is a northern beauty, quite at home among the ice and snow, like the arctic fox, hare or polar bear. Birch trees were among the first species to move northwards as the glaciers retreated after the last Ice Age, so they are one of Britain's oldest native species. The seeds scatter freely, blowing about like pale dust and sprouting almost wherever they land. The power of the pioneering pollen, so helpful for the tree's survival, is felt by millions every year, as it sets off the first hay fever of the season. The birch is packed with therapeutic benefits, nevertheless. As a source of betulinic acid, its antiretroviral and anti-inflammatory qualities are only beginning to be fully explored. The remedial reputation of the birch has never rested solely on its punitive habits. John Evelyn praised a contemporary birch concoction as 'a great opener' – and recommended it for pulmonary complaints and piles. The oil derived from birch trees is also supposed to be good for treating warts and eczema. Birch tea, made from an infusion of birch leaves in boiling water, may be a little bitter in the mouth, but apparently works wonders for gout, rheumatism and kidney stones.

Birch by-products are currently enjoying quite a revival, as the therapeutic benefits of birch water become more widely recognised. The liquid, which is regularly tapped from birch trees in Eastern Europe and Russia, is thought to lower cholesterol, reduce cellulite and boost the immune system. The sap of the birch, extracted through carefully aimed incisions under a mature branch, can also be boiled with honey, cloves and lemon rind, and then left to ferment until it turns into a very palatable wine. Birch sugar, which is made by boiling the sap, is a natural sweetener, but lower in calories and kinder to teeth than other kinds of sugar. In a memorable episode of the American Civil War, birch bark demonstrated its life-saving properties by providing General Richard Garnett's defeated troops with just enough birch sugar to keep them alive. Twenty years later, the course of their retreat was still visible from the trail of peeled birch trees. This was not the first time a hungry army had relied on birch trees: in 1814 the woods around Hamburg were devastated by Russian soldiers besieging the port and desperate for sap.

Birches grow on poor soil and tolerate extreme weather, and they thrive by seashores and highland streams alike. In countries where birch trees grow like weeds, almost every part has been turned to good effect. For the native peoples of North America, birch was as versatile as plastic: they turned it into bags, boxes and baskets, stitched together with threads of cedar-wood and decorated with patterns created by biting into the white surface. Above all, they used the timber for canoes – often known as 'birches'. Skilled birch builders would slide through the cold clear waters of the lakes, silent, stealthy and deadly efficient in their pursuit of fish and furs.

Despite being low in value as a timber tree, the birch has always provided a plentiful supply of firewood, which burns even when damp. From the earliest settlement of Britain and Ireland, birchwood was used for all manner of small items, including hoops and baskets, bowls and spoons, while the bark was shredded to make strings or twisted into ropes. All across northern Europe, birches have always been tapped for sap, while their bark was used for tanning. In Russia, the bark is distilled to make birch tar oil for proofing leather and deterring insects; when mixed with petroleum jelly it also works as a wood preservative. In Poland, bundles of birch twigs have been used for thatching cottages, while in Sweden, wooden houses are often insulated and waterproofed with birch bark. Nor is this a quaint detail from our collective rural past – one of the prize-winning entries in the World Architecture Festival of 2008 was the remarkable Restaurant Tusen at Ramundberget in Sweden, with its conical frame of dazzling birch poles, braced like a rocket launcher in the snow. The design not only protects diners from Arctic gales, but also proves that wooden buildings need no longer feel out of place among the futuristic creations of contemporary architects.

Since renewable materials have become resources of choice, what might once have seemed outdated methods have turned into state-of-the-art techniques. Green roofs, until recently regarded as mere picturesque features of folk museums, are now in demand across northern Europe. After the roof frame has been constructed, birch bark tiles are laid across the logs to form a base for the grass. A roof of this kind can last fifty years without major refurbishment. Modern eco-houses have bathrooms tiled with silver birch panels, and small shelves to act as soap dishes for Swedish silver-birch leaf soap. The house might even be completely lined with smooth panels of birch plywood and furnished with back-relieving bendy chairs, built from the timber of sustainable birch forests.

Canada is home to the 'paper birch' or *Betula papifyrus*, whose bark is even whiter and flakier than that of its relations in other continents. Although Canadian birch is best known for its capacity to slough a paper-like skin, bark of other birch species was similarly serviceable to the people of ancient China. After all, it does not take too much ingenuity to see the possibilities in the thin, white bark. When John Clare spotted some peeling birch fence poles one day, he immediately saw their potential as a supply of substitute paper. He soon discovered that 'one shred of bark round the tree would split into 10 or a dozen sheets', and, after a few experiments, found a way of ensuring that the ink would hold fast. For a man of very modest means, paper was prohibitively expensive, so the discovery of free materials for his writing was like a heaven-sent blessing. The generous birch trees seemed to be brushing away the obstacles of class and poverty and enabling Clare to reaffirm his true purpose in life.

The birch delivers its most salutary admonishments without any recourse to physical force. Often this tree seems to offer nothing more painful than a gentle invitation to come down to earth and begin again. In Nordmarka, the great wooded region north of Oslo, silver birches mark the way through the forest into one of the great secrets of the twenty-first century. For deep inside this snowy region is the Future Library, where a thousand spruce trees have been planted by the artist Katie Paterson. Leading writers from around the world are being invited to write stories, which will be kept locked away for a hundred years and then published on paper made from what are now no more than saplings. As it opens to reveal an unbuilt library, the mysterious birch challenges our demands for instant gratification, overnight celebrity and the triumph of the best-seller. None of us will be there to read Margaret Atwood's new work, but the Future Library is an expression of faith in a new generation.

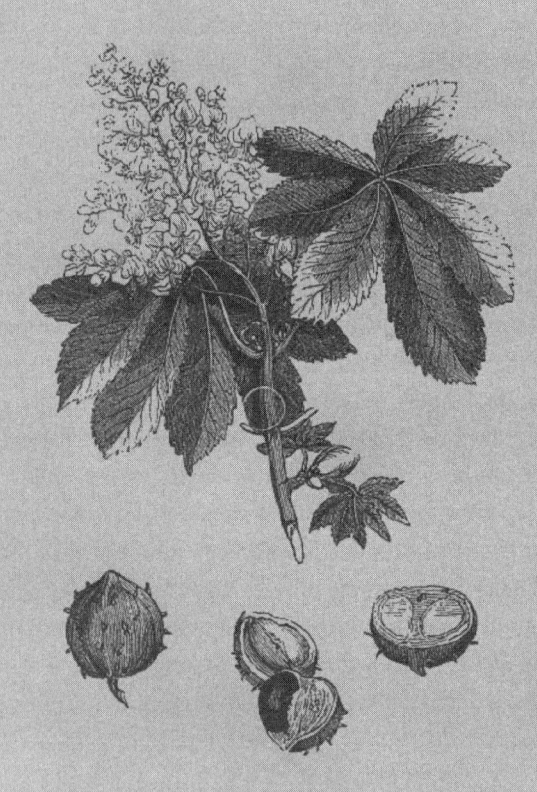

Horse Chestnut

THE biggest horse chestnut tree in Britain is at Hughendon, near High Wycombe. In December 2014, its girth had reached a gargantuan 7.33 metres and several of its branches had swollen to the size of respectable tree trunks. This is a tree of trees, a colossal, crinkled column with a copse on top. At three hundred years old, it is doing very well, since horse chestnuts rarely last longer than half that time. This means that when Benjamin Disraeli bought the estate at Hughendon in 1848, the horse chestnut tree, which now stands imperiously at the main gates, would already have been a large spreading sentry, very well placed to impress important visitors to the future prime minister's country mansion. Horse chestnuts thrive in the chalky soil of the Chilterns, where they make magnificent figures against the curves of the hills. Disraeli, sensitive about his own family origins, had a special fondness for a type of tree that had arrived from the Balkans only two and a half centuries before, but which had settled so successfully in England that it now seemed more native than many of the natives.

There is, after all, something wonderfully unapologetic about the horse chestnut tree. All through the winter, when deciduous trees should be lying dormant, its glistening buds just seem to bulge with pent-up energy, swelling with impatience until they can't wait a moment longer, and then, at the first touch of spring, they burst out of their sticky, constricting shell suits. Their frond fingers are finally free to unfurl and wave at the first fresh breeze. The huge, palmate leaves hold handfuls of sunlight before most trees have

GWEN RAVERAT, *HORSE CHESTNUTS AT GRANTCHESTER*

ventured out, their tooth-edged leaflets ready to take in the rays and the rain.

Horse chestnuts not only display the largest, greenest leaves, they also have another trick up their sleeves. As other trees are catching up on the colour of the season, the horse chestnut begins to send up its rocket sprays of blossom. By May they are covered in bloom, as creamy as champagne and just as bubbly. And it does not stop there, because, as the flower candles fade, the great green hands begin to sport spiky rings, which burst just as dramatically as the buds in spring. Strong winds in late September will bring the chestnuts crashing down, their lime-green cases splitting apart as they hit the ground. The nuts inside, so snug and round, merely seem to wink as they land, though some bounce free in glossy richness and readiness to go. The horse chestnut tree is packed with puzzles and private jokes, so confident of its own ability to steal the show that any questions about rights to residency seem to be missing the point.

The name of this familiar tree is itself something of a conundrum, whether it goes by the formal botanical label of *Aesculus hippocastanum* or the more familiar horse chestnut. What has it to do with horses? The pattern left by the leaf stalks after they have fallen does resemble a horseshoe, so this may have been the inspiration for the tree's name. Some claim that it derives from a Welsh word, *gwres*, meaning 'hot' or 'fierce', chosen because of the unpalatable nuts. It seems equally possible that the round shining nuts, peeping from the white shell, recall the eyes of a startled horse. The tree's American cousins, *Aesculus pavia*, *Aesculus glabra* and *Aesculus californica*, are, after all, commonly known as 'buckeyes', because the native peoples thought the round, brown chestnuts resembled the eyes of deer. (Ohio is known as the Buckeye State because of its chestnut trees, in a typically playful cascade of references whereby the state is named after the tree that is named after the deer that once lived there.)

Or perhaps the horse chestnut's name has nothing to do with horses' eyes and everything to do with the astonishing colour of the fresh horse chestnut? The rich, reddish brown nut, as first revealed when the shell splits asunder, is not unlike the glossy haunch of a healthy, chestnut horse. It may not be coincidental that the tree arrived in England around the same time as horses began to be described as chestnut. But which came first – the chestnut horse or the horse chestnut? The *Oxford English Dictionary*, usually such an invaluable authority, is surprisingly tentative on the derivation of 'horse chestnut', attributing its origin to an old belief that 'the people of the East' used the chestnuts for treating horses suffering from coughs and respiratory diseases. Since there is little evidence for the chestnut's value in equine medicine, the name may have been ironic. Unlike the wholesome fruit of the generous Spanish or sweet chestnut tree, *Castanea sativa*, the nuts of *Aesculus hippocastanum* are fit only for horses – and, indeed, not very good for them.

In comparison with the sweet Spanish chestnut, whose nuts are used not only for roasting on an open fire, but also for polenta and pudding, for marrons glacés and *montebianca*, for soup, stuffing and starch, and whose timber is stout enough to make pit-props, poles, furniture and roofing timber, the horse chestnut has relatively little to offer in practical terms. As a consequence, there are those who regard this amazing tree as something of a wastrel. Its timber is too soft for serious building purposes and it does not even burn particularly well. Its fruit is not especially nutritious – in fact the nuts would make you rather poorly. Even pigs, not renowned for their fastidiousness, turn up their snouts at a trough full of horse chestnuts. During the First World War, when food supplies were under threat, experiments involving a process of crushing, soaking and boiling the nuts resulted in an animal feed that was used for supplementing the diet of sheep and cows. The pigs were still unimpressed, but the chestnut meal helped to extend the meagre supplies until the war was over.

There *are* traditional uses for the horse chestnut – the bark, with its narcotic properties, was once deployed in the treatment of fevers, the nuts in cases of rheumatism and piles. This tree has long been regarded as a friend by those afflicted with even mild forms of arachnophobia, because the chestnuts are generally believed to repel spiders. Horse chestnut trees are not hospitable to spiders' webs, and many people hope that careful arrangements of conkers will keep spiders from coming inside when the weather turns cold. I have never been wholly convinced, not least since I tried to fortify a basement flat with battalions of stout chestnuts, only to find, to my great dismay, that the defences had been well and truly breached overnight. It seems that I am not the only one to harbour doubts about this item of folk wisdom. In the autumn of 2009, the Royal Society of Chemistry decided to put it to the test, issuing a challenge to anyone who could come up with conclusive evidence or, better still, a scientific explanation for the repellent properties of horse chestnuts. After some ingenious experiments involving conker obstacle courses and garden spiders, the theory was firmly disproved. No spider was remotely deterred by the sight, smell or touch of a horse chestnut.

But why should the value of a tree be measured by its utility? By the contribution it makes to human convenience? This is the tree with the fattest buds, the blousiest blooms, the broadest leaves, the spikiest seedpods, the shiniest nuts – so what more does it need? Horse chestnut trees are visually spectacular – extravagantly arrayed in every season. No wonder that they took western Europe by storm in the age of bright capes and silken waistcoats, lacy cuffs and ruffs. Some horse chestnuts, *Aesculus rubicunda*, have deep red blooms, to complement the rouge in the ladies' cheeks or the crimson ribbons in the gentlemen's pantaloons. These were the trees for the French aristocracy who loved to be entertained beneath their frothy flowers and shifting shade in the new gardens at Versailles. As things turned out, the horse chestnuts survived better than the aristocrats: Paris is still full of horse chestnut trees, spreading freely along the banks of the Seine or standing to attention, straight and square, all along the Champs Elysées. Beneath the Eiffel Tower is a monumental specimen. The tree was there long before Paris' most famous landmark, which acts as an enormously elaborate proscenium arch for this show-stealing veteran.

The flamboyant horse chestnut made quite an entrée into English parks and gardens, too, though its capacity to provide shade was not quite so much in demand as in Continental Europe. Sir Christopher Wren redesigned Bushy Park to create an enormous ceremonial avenue of horse chestnut trees, running from Teddington to the royal palace at Hampton Court, for William III. Queen Victoria took to making a ceremonial outing every May to see the fresh chestnut chandeliers, now fully established and at her service in their bright green and white livery. Her loyal subjects followed her lead, celebrating the blossom with brisk walks and brave picnics. Although the tradition declined after her death, it is currently enjoying a revival, as people turn out on 'Chestnut Sunday' to witness the spectacular show of blooms.

The eighteenth-century fashion for mass planting of these showy trees inspired Capability Brown to order 4,800 for a single estate at Tottenham in Wiltshire, which gives some idea of how highly esteemed these eastern Mediterranean incomers had become. In Oxford, huge horse chestnuts lean over the lake at Worcester College, dropping curling petals onto the smooth surface or offering a languid silhouette in the background of college ball and graduation photographs.

JAMES TISSOT, *THE PICNIC*

Another stands in the passage beside the Lamb and Flag pub, causing bicycles to swerve to avoid its substantial figure.

For many children growing up in modern cities, a sense of the natural world is rooted in horse chestnut trees. One of the abiding memories of W. B. Yeats' London boyhood was the great horse chestnut tree in the garden of the family home in Bedford Park, which eventually sprang into new life many years later as an image of wholeness and inner connection, the 'great rooted blossomer' in 'Among School Children'. His formative memories were reinforced on his return to Ireland by the grand horse chestnuts that flourish from Dublin to Galway. Many children absorb a sense of the seasonal cycle from the great green trees in parks and gardens with their bright white birthday cake candles.

In rural areas, mature horse chestnuts offer a gathering point for people from different generations, like the Morton Horse Chestnut in Derbyshire, which was designated one of the fifty Great British Trees for the Queen's Golden Jubilee. The familiar crimson blossom lights up the heart of the village of Cropredy in Oxfordshire each year, weeks before the annual music festival. These trees are traditional centres like the one celebrated in the old song 'Under the spreading chestnut tree', which even inspires the singers to imitate the welcoming branches.

The horse chestnut tree presides over pleasure and leisure, as the nineteenth-century artist James Tissot observed in his painting *Holyday* or, as it is more commonly known, *The Picnic*. The painting, which now hangs in Tate Britain, is set beside the lake in St John's Wood, home to the artist himself, but far better known as the home of Lord's Cricket Ground. The tree is the most arresting presence in the scene, which depicts a pair of young women, dressed in their finest, one of whom is pouring tea for a young man in a colourful red and yellow striped cricket cap and whites, stretching out beneath the wide-spread, green-gold, late September leaves.

Although September strikes the knell of the long summer holiday, for generations of children the dismaying prospect of going back to school was mitigated, until very recently, by the imminence of the conker season. For any pupils lucky enough to have a horse chestnut growing in the school grounds, this was the moment to hurl sticks and shoes into the lower branches in the hope of bringing down the mace-like cases.

All you need to play conkers is a horse chestnut threaded onto the end of a knotted shoe-lace, a similarly equipped opponent and a destructive inclination. An especially powerful horse chestnut can rack up quite a score, as one after another puny challenger is smashed to smithereens. Those seriously bent on victory have been known to resort to somewhat underhand measures, such as soaking their champion horse chestnut in vinegar or baking it in a hot oven to maximise the strength. These ploys seem quite innocent, however, in comparison to that of one of my uncles who, I'm told, was so determined to be hailed the conquering hero that he crafted a conker from solid wood and polished it for victory. Even the toughest horse chestnut stood no chance against this legendary smasher, though the

BOYS PLAYING CONKERS IN THE 1950S

sense of triumph may have felt more hollow than the champion conker. For children less bent on world domination, horse chestnuts still provide plenty of scope. They lend themselves as elliptical marbles and, with a row of thimbles, to diminutive coconut shies. A few well-placed pins and a reel of cotton can transform horse chestnuts into doll's house furniture, which begins by resembling polished mahogany, but gradually fades to duller, more wrinkled forms.

Playing conkers may seem one of those quaint pastimes, enjoyed since time immemorial, but in fact it is another of the many traditions invented by the Victorians. It is also one that may well have had its day. So noticeable has the recent decline in this seasonal activity been that the Health and Safety Executive felt the need to issue a denial that they had banned the game from school playgrounds. It seems that the end of conker matches may have less to do with safety-conscious head teachers than with the more compelling attraction of mobile phones.

If conkers no longer command the interest of the nation's children, their continuing appeal is abundantly evident at the World Conker Championships, which take place every year in the

Northamptonshire village of Ashton, attracting contestants from surprisingly far-flung parts. As an international sport, conkers requires rules of engagement and so there are strict guidelines on the length of the lace, the manner of knotting it and the number of strikes. Those who ultimately seize a glorious victory are adorned with chains of glossy horse chestnuts and colourful, conker-studded crowns. As a rustic counterpart of London's Pearly Kings and Queens, the Conker Monarchs could easily be mistaken for something hallowed by ancient tradition, but it all seems to have started in 1965.

It is not only the fruit that raises competitive spirits. Some countries are so proud of their horse chestnut trees that they enter them in international competitions. Belgium's candidate for the European Tree of the Year 2015 was the fine veteran, known as the 'Nail Tree', which clings to a bank at Voeren on the Dutch border in Limburg. This splendid horse chestnut, whose bark is flaking with age, has long been revered as a healing tree. In years gone by, people who were ill would place a nail against their ailment before hammering it into the tree, in the hope that the tree would take away their pain. Attached to the trunk is a crucifix, to link the horse chestnut tree with Christ's suffering and the wooden cross.

After the Second World War, the fame of the Nail Tree was eclipsed by another horse chestnut that grew for many years in Amsterdam. This was the tree that Anne Frank could see through the small window of the secret annexe where she and her family hid from the Nazi occupation force. The diary she kept throughout these unimaginable months of enforced captivity and unremitting anxiety records the seasonal miracle performed by the horse chestnut tree. On 18 April 1944, Anne wrote in her diary that the chestnut tree was 'already quite greenish, and you can even see little blooms here and there'. By 13 May, the day after her father's birthday, she noted that the sun was shining 'as it has never shone before in 1944' and the horse chestnut was 'in full bloom, thickly covered with leaves and much more beautiful than last year'. Within three months, the Frank family had been betrayed to the Nazis, arrested and sent off to successive concentration camps. Anne and her sister Margot died in Bergen-Belsen, only weeks before the war ended. The tree that had made her so happy, simply by being there, continued to live nevertheless and sent out its bright leaves and dazzling candles every spring.

Eventually, however, Anne Frank's horse chestnut began to show its age. By the turn of the millennium, riddled with fungus and infested with insects, the tree was becoming a threat to the very visitors who came to express their admiration for its life-affirming powers. In 2007 a felling order was issued, but the public outcry was so intense that the tree was reprieved and efforts made to preserve its ailing trunk and failing boughs. A severe gale in 2010, however, proved more than it could withstand and the old horse chestnut tree was blown down. Saplings from Anne Frank's tree have since been planted across the world in remembrance of the hope it gave to one of the war's most famous victims and to lift the hearts of later generations.

The indignation over the decline of this particular tree is not just a testament to the memory of Anne Frank – it also reveals much about the essential meaning of the horse chestnut. This seemingly irrepressible tree stands for health and energy and life itself. It is so tenacious that even after the leaves have dropped, they refuse to lie down flat. Their strong veins meet along the copper-coloured spine as if invigorated by the falling temperatures. Under a hard frost, they curl into quill pens, poised to record some marvellous secret of their own.

This is what makes a sick horse chestnut tree so disturbing. For a tree that has always seemed to be overflowing with health, an arboreal disease is especially distressing. In recent decades, these effervescent giants have been in danger of being laid low by the larvae of a tiny moth burrowing into those lovely leaves. The leaf miner has the power to strip the gloss from the summer foliage, leaving it limp and brown. The sight of the great Parisian horse chestnuts withering away at the height of summer casts a serious dampener over the holiday atmosphere. It is one thing for this tree to seem eager for the spring, quite another for it to reveal signs of an accelerated autumn.

More worrying still is the bacterial infection spreading through the horse chestnut population of Britain, which makes the bark bleed and eventually kills the tree. The dark, oozing substance is the tree's defence against bleeding canker, which is an infection caused by the bacterium *Pseudomonas syringae* pv *Aesculi*, and the plant pathogen *Phytophthora*. There is no chemical treatment as yet. The extent of the infection can be seen by stripping off the weeping wood around the wound to reveal the bruise-coloured spread on the inner bark. In some cases, the infection is contained in a relatively small area, enabling the tree to carry on growing, but if the disease reaches round the entire trunk, it stands little chance of survival. Already swathes of Irish horse chestnuts have been struck by bleeding canker.

Suddenly, the value of the horse chestnut has become more obvious than ever before. So far the depredations of the leaf miner have turned out to be more of a passing irritation than a chronic condition. There is a chance that bleeding canker may be a health scare rather than a death sentence, since faking its own death seems just the sort of trick that the horse chestnut might pull off – but we probably should not bank on it.

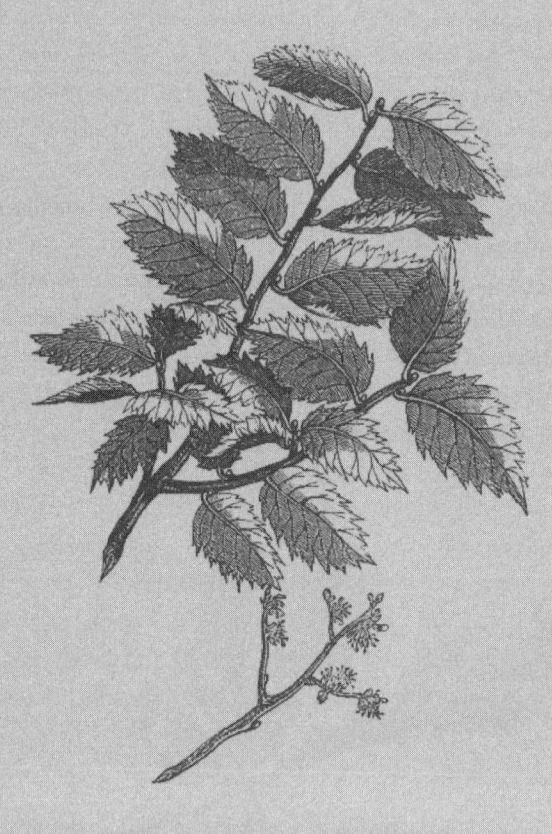

ELM

IN the corner of the Market Square in Stony Stratford, a tree has been put behind bars. Since it is a young, healthy oak tree, growing straight and tall, the cage seems to have been installed as a protective measure, forestalling any unwelcome attention from people who might think it fun to demolish a sapling. A closer look reveals, however, that the round cage is surrounded by a further line of defensive railings, which seems a somewhat overcautious measure in the square of a small, market town. Inside, a little plaque sheds light on the curious fencing, explaining that the oak tree stands on the site of an old elm where John Wesley had once preached. Many residents of Stony Stratford still remember the oak's venerable predecessor, probably more as a stump than as a tree, though still sprouting a rash of nettle-shaped leaves each spring as a gesture of defiance. Stony Stratford was once a staging post where travellers broke their long journeys from London to Holyhead, and for many years the forlorn, headless tree stood with pathetic irony right outside an old coaching inn called the Crown.

The hollow relic was all that remained of the magnificent elm where Wesley once held the open-air prayer meetings that challenged the regular goings on of the old parish church on the other side of the square. The plaque commemorating his impromptu pulpit was commissioned in 1950 by a member of the local Methodist church, but at that time the elm was even larger and more sprawling than it had been in Wesley's day. Within thirty years, Dutch elm disease had reduced it to a mere wreckage of its former self, and then a few stray, smouldering cigarette ends finished it off. The railings were erected to protect the old tree, but the effect was rather of a maimed prisoner who had been in solitary confinement so long that no one paused to wonder why. The elm stump was left to stand, nevertheless, as a monument to both Wesley's mission and to all the other English elm trees that once flourished throughout the country. Like them, Wesley's elm is now seen no longer.

From essential to endangered to extinct: the elm has undergone an utter transformation in modern consciousness. The tree that was once a fundamental strand in the very texture of the British landscape was more or less expunged in a single decade. And now the few remaining places where elms survive to maturity have become sanctuaries, defined by the chance survival of what was once commonplace. When London's Great Trees were being catalogued after the great storm of October 1987, which had turned so many tree-lined streets into demolition sites, the Marylebone Elm was the last of its kind left in the Borough of Westminster. The Brighton elms are part of the city's colourful culture, now as eccentric, in their way, as the Royal Pavilion. As is the way with sudden extinctions, the ordinary becomes extraordinary simply by virtue of still being there.

Through much of England and Wales, the sight of a mature elm is largely a thing of the past. They are instantly recognisable from almost any older landscape painting: defining features of 'Constable country', distinctive presences in Turner's sketches of the tree-lined Thames, soft silhouettes in the background of George Stubbs' *Reapers*. They are there in postcards of English villages and preserved in films and photographs taken before the 1970s: those tall, often slightly top-heavy trees with clumpy foliage like rough, green, unspun wool, wrapped around a very solid spindle. The leaves are small and serrated, unfurling early to catch the first spring sunlight, but slightly rough on the surface. Elms sprang up along East Anglian lanes and rivers and they shaded the longboats that passed slowly down the canals of the Midlands. They formed lofty walkways in ancient cities, like the New Walk along the Ouse in

GEORGE STUBBS, *REAPERS*

Georgian York or the broad walk through Christ Church meadow to the Isis in Oxford. An elm of twenty years' growth can survive transplantation, so they were ideal for creating elegant geometric patterns in eighteenth-century landscape gardens: for the Great Bow surrounding the Round Pond in Kensington Gardens; or for the Grand Avenue at Stowe which rose steadily for more than a mile towards the triumphal Corinthian arch on top of the hill. In a breeze, elm leaves flutter prettily to soften straight, faintly militaristic lines, but in a stronger wind, a row of these trees would toss and turn like horses freed after a day's hard work. This is no longer a familiar sight. The few English elms that remain are the last of their race – living links to a family so vigorous that they flourished from Cornwall to Cumberland, from Kent to Carnarvon. These trees inevitably provoke very different thoughts from other common native species, and their traditional meanings are now indelibly tinted by the fact of absence.

Since the devastation of the last century, the very word 'elm' has been inextricably linked with 'disease'. Although there had been a severe outbreak of Dutch elm disease in the 1920s, it was not on the scale of the arboreal epidemic that raged through the 1970s, leaving the country permanently altered as Margaret Thatcher swept to power. The new, more virulent, strain of Dutch elm disease was only detected for the first time in the late 1960s, but in little more than a decade it had laid waste to the elm population of Britain, destroying some twenty-five million trees. The unstoppable plague made a mockery of plant scientists and conservationists, who seemed powerless to halt its rapacious progress. Once stricken, nothing could be done to prevent a row of huge, healthy elms sickening one by one until nothing remained but shivering skeletons. Village greens and grand avenues, ancient veterans and ordinary hedgerows alike succumbed to a disease that showed no respect for age, location or historic status.

Dutch elm disease was disturbing, too, because of the proximity of the apparent source – Holland, the land of milk and cheeses, tiles and tulips; in other words, the quintessence of wholesomeness, but now supposedly the launch pad of a deadly invader. The whiff of treachery made the invisible menace even more unsettling. In fact Dutch elm disease did not come from Holland at all, but arrived from the other side of the Atlantic, probably on infected wooden poles. The name came from the research into the disease, which was spearheaded by Dutch scientists in the 1920s. At the heart of the matter is a deadly fungus, *Ophiostomi novo-ulmi* (formerly known as *Ceratocystis ulmi*), which causes blockages in the tree's internal irrigation system, preventing water from reaching the crown. The tree withers from the head downwards, and once the fungus has crept inside the roots, it is doomed. Little, brown-winged beetles, *Scolytus scolytus*, carry the fungal spores from tree to tree, as they alight on the trunk and bore down inside the deeply furrowed bark to lay their eggs. The first signs of illness are the stricken boughs protruding above the canopy like skeletal arms reaching for the sky, the dimpling lumps and bumps on the leaves and then the tell-tale dark stains inside the trunk; but it can take a year or more for the entire tree to succumb. The decline of the elm also sounded the death knell for other species dependent on its generous spread. The large tortoiseshell butterfly with its bright tiger-stripings was once among the commonest British insects, but is now extinct. The fate of the white-letter hairstreak butterfly, whose sole source of sustenance is elm foliage, hangs precariously in the balance among the leaves of some of the country's few surviving veteran elms in Preston Park in Brighton.

Elms thrive on sea air, and in Brighton they have filled every park and city square ever since the Prince Regent put his not inconsiderable weight behind a substantial elm-planting programme in the early nineteenth century. In the 1970s, the city council, acting on advice from their local arborealist, Rob Greenland, decided to take drastic measures to save their signature trees. The South Downs form a natural barrier between Brighton and the rest of England, helping to create a micro-climate and a potential pocket of resistance against even Dutch elm disease. A desperate programme of felling infected trees began to ensure the safety of their healthy companions and to create a cordon sanitaire across the Downs. Brighton is now home to the National Elm Collection, which stands proudly along The Level in the heart of the city. Although the population is severely depleted since its nineteenth-century heyday, when over a thousand elms were planted, it is one of the only places in England where you can see a group of fine mature trees. Just outside the city, the University of Sussex has its own large specimens, which long predate the campus and the students who hurry from lectures to lunch. Although the story of Brighton's elms is one of heroic triumph against the odds, they are still in constant danger: despite the care of the local disease officer and the constant vigilance of many residents throughout the summer months when attack is most likely, the city still lost about thirty mature elms in 2014. With the elms the focus of such passionate communal effort, the fall of a single tree is a local tragedy.

Mature elms have effectively become extinct in many areas, but it is still possible to find younger members of the family, shooting up as confidently as any of their ancestors. The devastating disease appears to spare the saplings, striking only when the trees are about fifteen years old. Whether this is because the death-heralding bark beetles limit their attacks to more substantial trunks or because the new shoots spring up from roots already infected with the fungus, these young trees have no hope of reaching full maturity. Elms generally propagate by suckering, which is why they used to grow up in great lines and clumps from a single tap root, as 'a brotherhood of elms', but separate trees can also fuse underground, linking roots companionably. Early settlements were often surrounded by elms, which acted as a screen against harsh weather or a natural camouflage from marauding invaders. Now their natural habits of camaraderie and protectiveness are also their fatal flaws.

The elm, then, is no longer the large bushy native that once stood side by side with poplars, oaks and ashes, but a slim, youthful upstart of a tree, doomed to premature death, as it loses life from the crown downwards, finally left to stand like a blasted wreck until it is felled or fallen. The elm means lost strength, disappearance and nostalgia, or, perhaps even more poignant, it suggests youthful hope blasted. A modern elm tree is an arboreal fragment inspiring thoughts of vanished wholeness or of a future immensity that will never come into being.

There have been attempts to replenish the population by planting what appear to be disease-resistant elms, often imported from abroad. Chinese and Japanese elms have proved largely immune to Dutch elm disease and some American elms are much less susceptible than those in Britain. There are even varieties of native elm that have not fared as badly, including the narrow-leaved Huntingdon elm and the wych elm, or Scottish elm, which can still be found quite widely in the far north. It is the English elm that has been so hard hit, and whose sudden catastrophic defeat has completely changed our understanding of this once ubiquitous species. Glimpses of its unassuming but fundamental role in English culture are made possible by paintings, descriptive writing, stories, artefacts, historical records and botanical studies, but it all has the air of being once removed. Fortunately, the elms of East Anglia were the subject of an exhaustive study conducted between 1955 and 1976 by R. H. Richens, whose brilliant book, *The Elm*, published just after many of its subjects had suddenly perished, combines scientific obsessiveness with a powerful sense of personal bereavement.

In the poetry of earlier rural writers such as John Clare, we can sense the sheer ordinariness of the elm in passing references to a tree that provides welcome shade to sheep shearers or those working in hot harvest fields. Francis Kilvert's diary of the life of a later nineteenth-century clergyman in the Welsh Borders, too, is packed with references to the great elms that shaded the lanes and the local churchyards. Parish records and county histories confirm that ancient elms served as meeting places and boundary trees, like the Tubney Elm, on the boundary of Fyfield in Oxfordshire, whose girth of thirty-six feet made it an obvious local landmark. At Sigglesthorne near Hornsea, in the East Riding of Yorkshire, the route to the village well was marked by three ancient elms, called the Bass, Alto and Tenor, under whose sonorous branches people would sit and put the world to rights, before heading home with their jugs of water. An ancient elm tree could live for more than seven hundred years, gradually becoming a permanent feature in the local land and mindscape. The Crawley Elm, for example, had attained a height of seventy feet and a girth nearly as large when Jacob Strutt went to visit it. The grandparents who were supervising the children who clambered over its massive roots and lower branches could remember their own grandparents recounting tales of *their* childhood games in the elm. This was more a much-loved family member than a tree. E. M. Forster captures something of the sense of spiritual continuity embodied in the elm tree in his novel *Howards End*, where an old wych elm in the garden stands for a kind of understanding utterly at odds with the practical, progressive attitudes of the family who regard themselves as its owners.

Strutt's celebratory survey not only included his drawings of the great trees of Britain, but also the thoughts they inspired when he

THE CRAWLEY ELM FROM JACOB STRUTT'S *SYLVA BRITANNICA*

was taking their portraits. Elms were clearly giving rise to anxiety even then, because of their susceptibility to disease. Strutt noted with some concern the recent disappearance of elm trees from St James's Park, for example. The great elms seemed to offer their own retort to such fears about the future through the vigour immediately evident in the early buds, abundant foliage and astonishing longevity. Whatever the effect of pests and diseases, there was no real sense before the 1970s that the elm would not continue to rise triumphant over British soil. Among the first trees

WILLIAM HENRY FOX TALBOT, *ELM IN WINTER*, CA. 1845

to have its portrait taken by a photographer was the Great Elm at Lacock, immortalised in William Fox Talbot's extraordinary image from the 1840s. Its leafless silhouette has a strange, ghostly air as it towers above the soft sepia undergrowth like some otherworldly warrior emerging from the mist. This is a giant of the open fields, magnificent, awe-inspiring and seemingly immune to the ravages of time.

It was Thomas Hardy who really captured the intense relationship between ordinary people and the English elm. In his most tree-lined novel, *The Woodlanders*, an elm plays a crucial role. Mr South, the ailing older resident of Little Hintock, spends his days worrying about the large tree outside his cottage. So obsessed is he with the fear of it falling and crushing the family home that he becomes too ill to go out. All attempts to lift the overhanging gloom, including the traditional practice of elm stripping, whereby the branches are thinned to let the light through, fail; and so, when a newly trained doctor from London arrives on the scene, he makes what seems a logical decision. The tree must come down. As the tree is felled, the physical shadow lifts from Mr South's life, but the shock of this sudden transformation proves far more devastating: by the next day, Mr South is dead. Hardy understood better than most the complicated intimacy of people and trees – over the years Mr South's entire sense of himself and his view of the world have become dependent on the all-powerful elm, whose unmistakable presence is both terrifying and yet strangely life-affirming. To lose it is to lose himself – he cannot survive its removal.

Hardy's choice of the elm for this parable of neurotic psychology was typically astute. Elm trees tend to have shallow roots, and so they are especially prone to crashing down in gale-force winds. Many of the elms that lived through Dutch elm disease in the 1970s were flattened by the great storm of 1987. It is a hazard even now: one of Suffolk's rare survivors, an ancient 250-foot giant elm, crashed through the roofs of nearby bungalows in Worlingham, early one day in May 2015. Elm trees are also given to dropping large limbs without warning, as a New York social worker found to her great misfortune on a warm summer evening in July 2007, when she was out with her dog in Stuyvesant Square Park, Manhattan. She sat down on a bench beneath a tall elm and, minutes later, was crushed by the sudden descent of a huge branch. There are many similar instances of elm limbs falling without warning: Gilbert White refers to the collapse of a huge elm branch at Selborne in 1703, which was evidently still a topic for local conversation eight decades later. Mr South's fear of the elm by his cottage in Little Hintock, then, has an entirely rational basis. Hardy shows nevertheless that once anxiety takes root, it will send up suckers to infect the entire mind.

Hardy was undoubtedly aware, too, of the elm's reputation as the coffin tree. Until the 1970s, there was always a ready supply of elm-wood in England. The planks were strong and unusually durable even in extreme conditions. Elm branches were often hollowed out to form pipes, because they were much less prone to rotting than other timbers and so early water mains systems often relied on these trees. Elm-wood was in demand for the paddles of steamers and watermills, and shipbuilders also used it for keels and gun carriages in the sides of warships, and indeed for anything designed for constant exposure to salt water. Great spars of elm were used for the foundations of bridges, while elm-planks offered sturdy weatherboards for barns and farmhouses. When it came to encasing bodies under six feet of damp earth, elm was an obvious choice. Thomas Gray, observing the 'rugged elms' that surrounded the graves in his 'Elegy Written in a Country Churchyard', probably took for granted that everyone knew these trees as the source of planks for coffins.

Since Gray's churchyard was at Stoke Poges in South Buckinghamshire, it is likely that some of those whose hidden lives are lamented in the poem were furniture makers who would, when alive, have sized up the rugged elms as potential timber. Elm was much in demand in the Chilterns, which was a natural – and national – centre for chair manufacturing. The classic Windsor chair is made from a bentwood wheel-back of ash or oak, with turned spindles of beech and an elm seat. A good block of elm-wood, with its interlocking grain, was strong enough to resist the steady onslaught of the adze and the travisher without splitting in two. The skilled craftsmen who laboriously scooped and sanded and polished the wooden block, gradually transforming it into a smooth, undulating platform fit for the human form, were known, unsurprisingly, as 'bottomers' (to set them apart from the 'bodgers' who made the other parts of the chair). A polished elm seat is often so carefully crafted that the lines seem to spread from the central rise like peaty water rippling away from a smooth stone. There is no need for a cushion if the bottomer got things right. The comfort-giving and water-resistant qualities of elm-wood meant that it was also the best choice for lavatory seats in the days when they were fashioned from wood rather than being stamped from plastic moulds.

Though elm was in many ways the most domestic of woods, the tree still often carried faintly disturbing associations. Lurid legends concerning elms and graves have lingered long in local minds. Maude's Elm, which grew in the village of Swindon, now largely absorbed by modern Cheltenham, was supposed to have grown from the wooden spike thrust into Maude Bowen's heart when she was buried at the village crossroads, because at that (unspecified) time, suicides were denied Christian burial. The story goes like this: Maude was found lying naked in the river, having apparently drowned on her way back from delivering hand-spun wool on her regular trip into town. The body of her uncle was also found, with a fatal arrow wound in his chest. Both deaths seemed inexplicable, but Maude was thought to have drowned herself. Her mother, inconsol-able with grief, would sit for hours beside her daughter's grave where the young elm tree was beginning to shoot up. Eventually the local squire had had enough of this and came to move her on but, as he rough-handled the helpless woman, one of his men was shot by an arrow. Maude's mother was arrested, tried for witchcraft and sentenced to be burned at the site of the grave. In the horrific events that followed, she was tied to the elm and surrounded by faggots. As the pyre was set alight, the squire looked on and jeered, until an arrow came from nowhere, pitching him into the flames.

Many years passed before the strange tangle of events unravelled, following the arrival of an elderly man at the Bowens' cottage. He confessed that he was responsible for killing the squire, his servant and Maude's uncle, because he had been in love with Maude and

BIRKET FOSTER'S ILLUSTRATION OF 'THE RUINED COTTAGE'

witnessed her rape by the squire and her own uncle. Whether or not there is any truth in the tale, Maude's Elm evidently survived the fire to flourish for centuries, widely regarded as a symbol of unjust suffering at the hands of those in positions of power. Although the great tree marked a place of violence, it also preserved the names of those cast out by society, forming an eloquent, living memorial at the unmarked grave.

Elms often stood for more than initially meets the eye. In Wordsworth's moving story 'The Ruined Cottage', a group of lofty elms growing from the same tap root provides a recurrent motif of life and community within a very human narrative of desertion, despair and death. The healthy trees continue to loom over the derelict cottage, long after the death of the last human tenants, with an enigmatic presence that speaks to those who knew and loved them. It is difficult to grasp the significance of the elms, once the marker of a family home and a meeting place for travellers, now perhaps serving as a memorial or, perhaps, just as a reminder of nature's 'calm oblivious tendencies'. The poem is a dialogue between a wise old man and a younger traveller, and the enormous elms tower over both, sheltering and casting shadows.

In Matthew Arnold's elegy for his friend and fellow poet, Arthur Hugh Clough, a 'single elm-tree bright /Against the west' plays a key role in the grieving process. As the bereft speaker tries to relive the heady days of his youthful friendship in memory, he recalls the pledge made at the foot of the tree:

> while it stood, we said,
> Our friend, the Gipsy-Scholar, was not dead;
> While the tree lived, he in these fields lived on.

The old elm had been there long before it was discovered by Arnold and Clough, and it would continue to thrive long after they were gone. As such it seemed the guardian of values embodied in their shared world of ideals – of a world untainted by economic imperatives, market forces or the more destructive tendencies of modernity. In his grief, Arnold found comfort in revisiting the elm, despite his own lacerating loneliness:

> Despair I will not, while I yet descry
> 'Neath the mild canopy of English air
> That lonely tree against the western sky.

The very isolation of the tree and the evening tones of the western sky nevertheless suggest some underlying doubts. Arnold may be turning to the tree for reassurance, but something about it hints at unspoken vulnerabilities and darkening horizons.

Elms have now become incontrovertible figures of loss. In the Royal Botanic Garden in Edinburgh there is another circle of bars with the remains of an elm inside. The Aeolian Pavilion is a memorial to the species, but instead of relying on a small plaque like the one in Stony Stratford, this contains a more tangible monument: a large Celtic harp crafted from the old wych elm that finally yielded to Dutch elm disease in 2003. The tree is gone and the garden emptier, as the wind plays over the Ossianic harp strings, creating a plaintive requiem.

What is so moving about the elm tree is the realisation that its traditional meanings were not, after all, out of keeping with its fate. For centuries the elm oversaw the human lives and deaths that took place beneath its boughs, even, on occasion, playing a causal role. The timber of this tree accompanied people to their final resting place, sharing their damp graves. But, at last the elm itself was laid low. When Byron was at Harrow, he used to spend time sitting under the old elm in the graveyard, reflecting on the passage of time and the transience of joy. In the elegiac poem inspired by the melancholy spot, he imagined the branches of the drooping elm whispering, 'Take, while thou canst, a lingering, last farewell'. When he wrote the lines, Byron was lamenting the departure of his own childhood. Only a few years later, when his daughter Allegra died at the age of five, she was buried at Harrow, beneath her father's favourite elm. The lines of Byron's youthful, melancholy poem accordingly became tinged with a much deeper, more specific grief. Now this great tree, like so many of its family, has gone as well, and so Byron's lines have become as prophetic as they were nostalgic. The fanciful words of his favourite boyhood tree now serve as an epitaph to the English elm.

WILLOW

THE willow's long, elegant leaf, divided by a pale central stem, is slightly upturned at the tip, rather like the Mona Lisa's smile. A mature willow tree is a mass of these light, half-smiling lips, murmuring softly with every breeze that blows, whispering at the slightest breath, quivering a little even when the air is still. When the wind is stronger, they begin to chatter and laugh as the breeze ripples through an unravelling scale of twigs. By the autumn, though brown and thinner now, they are still rattling out a last discordant chorus, whipped up into a frenzy of beating sticks, before being swept away in the blast. From their first tentative buddings to their final frosty defeat, willow sprays are in perpetual motion, filling the air with noises. But what is the willow saying?

What we might expect to hear are gentle sighs or inconsolable sobs. The willow has long been known as the tree of loss – ever since the exiled Israelites hung their harps in its drooping boughs by the waters of Babylon. In the folk-rock vogue of the 1970s, Harry Nilsson implored audiences to listen to the wailing of the willows while Steeleye Span imprinted the indelible beat of a willow-clad hat and a true love far, far, away on the minds of a vinyl generation. In between the Psalms and the seventies comes a sorry procession of abandoned lovers and broken hearts. From old folk ballads to jazz standards, the song of the willow is a sad one. Shakespeare did much to strengthen the willow's melancholy associations: in *The Merchant of Venice*, Lorenzo imagines Dido left holding only a willow bough as Aeneas sails away; in *Hamlet*, Gertrude paints a moving picture

AGNES MILLER PARKER, WILLOW CATKINS

of Ophelia slipping into her watery grave under the willow aslant the brook; but most painful of all is Desdemona's rendition of the 'Green Willow' song on the night of her murder. For W. B. Yeats, lost love was not just for the girls: in 'Down by the Salley Gardens', the willow is once again a site for trysting and tristesse. (Among the willow's alternative names is the salley, sally or sallow, or, in Irish, *saileach*, and all are closely related to the botanical Latin name, *Salix*.) In tender songs like this, the soft, swaying sprays of the willow seem to whisper sweet secrets before love slips away.

The long tradition of heart-rending 'Willow' songs was eventually parodied by Gilbert and Sullivan, who took the standard tree by a river as the perch for a little tom-tit singing 'Willow, titwillow, titwillow'. So plaintive is his refrain that the listener, Ko-Ko, is moved to enquire, 'Is it weakness of intellect, birdie, I cried, / Or a rather large worm in your little inside?' The answer, of course, is 'Willow, titwillow titwillow'. The song is a light-hearted piece of emotional blackmail by Ko-Ko, whose advances are being spurned, with the little bird who 'sobbed' and 'sighed' before plunging himself into 'the billowy wave', offering a cautionary model for those disappointed in love and those whose unrelenting hearts might drive their admirers to desperate measures. The willow is the tree of blighted affection and suicidal despair: something about its very nature, it seems, makes everyone want to weep. And this may seem natural enough for a tree with such stupendously pendulous branches and dripping extremities. The problem with leaping to this easy explanation, though, is that many of the folk songs were flourishing in Britain long before the arrival of the weeping willow. The now ubiquitous weeping variety was entirely unfamiliar to Shakespeare or the singers of traditional willow folk songs, which is why imagining Ophelia drifting away beneath long, trailing boughs, though very decorative, is distinctly ahistorical.

Salix babylonica (as it was named by Linnaeus in honour of those famous biblical willows, even though the tree actually originates in China), did not become established in Britain until the eighteenth century. Legend has it that the poet and keen gardener, Alexander Pope, was responsible for its introduction. His neighbour at Twickenham, Henrietta Howard, the Countess of Suffolk, whose magnificent Palladian mansion on the Thames owed its existence to her lover, George II, received a gift of figs from Turkey. Pope is said to have eyed the packaging and asked for a twig from the exotic

ALEXANDER POPE'S VILLA AT TWICKENHAM

basket. He then planted it in the garden of his own villa, a little further along the river, where it grew into a fine weeping willow. It is much more likely that if Pope did plant a weeping willow by the Thames, the original cutting came from his landlord, Thomas Vernon, who made his fortune from the lucrative trade with the Levant and whose botanical interests probably inspired the import of such a strikingly beautiful specimen tree.

Pope did his bit to help establish the weeping willow in Britain by sending some willow twigs off to friends in Bath shortly before his death, but his real contribution to willow culture was inadvertent and posthumous. 'Pope's willow' was largely a Romantic invention, featuring in paintings and drawings as an elegant subject in the foreground of his sloping riverside garden only some years *after* his death. With the demolition of the poet's home in 1807, the weeping willow was chopped down, leaving little more than a stump at the entrance of his famous grotto. The tree survived in small wooden trinkets and pieces of jewellery, to be treasured by poetry lovers like fragments of the True Cross. Guidebooks to the Thames and river tours went on lamenting the destruction of one of England's sacred literary sites for many years, transforming Pope's willow into an emblem of loss – a sentimental memory for a very unsentimental poet.

By the end of the eighteenth century, the distinctive outline of the cascading tree was widely recognised in Britain – not least because of the popularity of the blue and white Willow pattern that appeared on tea sets and dinner services everywhere. The luminous blue design of the pagoda with its walled, Chinese lakeside garden, a boat in the background and a pair of birds flying high across the white expanse above, is still in production, issuing subliminal willow messages beneath soup or steak or salad. The plates depict an Eastern tale of young love thwarted by a tyrannical father and a fiercely vindictive aristocratic suitor. The mandarin's daughter, rebelling against her father's choice of a wealthy, powerful prospective husband, elopes with her rather less socially elevated lover on the very eve of the wedding, just as the willow blossoms are about

WILLOW PATTERN PLATE

to fall. The young couple escape from the palace, across a bridge to a secret island, where they remain in blissful togetherness until the rejected suitor discovers their refuge and sends in his troops. After their death, the young couple transcend the mortal world in the shape of lovebirds. At the centre of the ceramic design, dominating the palace garden, the bridge and the boat, stands the most sinuous willow, with its bright blue fronds waving the lovers on their way.

Although the story fitted the contemporary fashion for all things oriental, it seems to have been an entirely English invention. The ingenious ceramics artist Thomas Minton had a very shrewd idea of what China meant to his customers – and of what might sell. Willow pottery is an early marketing success story, showing how preconceived ideas and a romantic narrative can be carefully exploited for commercial purposes. The massive popularity of the blue and white plates also did much to strengthen the associations of weeping willows with sorrowful tales of love and loss. As the Chinese-style colouring and the detail of the pagoda and sampan make clear, the weeping willow was still widely recognised in late eighteenth-century Britain as an exotic eastern tree. The newly established *Salix babylonica* provided perfect stock for grafting on traditional native associations between the willow tree and lamentation. With the arrival of this eastern variety, older ideas about weeping and willows were united in a living, physical reality.

The willow of the folk songs must have belonged to the many indigenous varieties of the *Salix* – such as the white willow, the crack willow, the goat willow, the grey willow, the bay willow or the common osier. All have been thriving in Britain for many centuries, but not one of them is a weeper. Many willows, in fact, look rather like Struwwelpeter, with their springy branches bursting from a tufty crown and the sprays of smaller twigs silhouetted against the sky. When a person is described as 'willowy' it usually suggests a light, youthful, sylphlike physique, but many willow trees are only too prone to middle-age spread. In fact, they grow so rapidly that the trunks of older trees are in danger of tearing themselves in two because of the great weight they have to bear. The crack willow, *Salix fragilis*, all too often lives up to its common name by splitting loudly down the middle. An unexpected summer storm can cleave a spreading veteran in two as surely as a well-aimed axe will split a log. Mature trees strain under their own potentially lethal abundance, as if their entire existence is one long strongman competition. As soon as the wind gets up, the fat old willow in our garden begins to twist and growl as if to remind us that a creak can very quickly become a crack.

The problem is eased by frequent pollarding, the removal of branches taking the weight from the crown. This traditional practice also provides regular supplies of slender olive-coloured poles, which can be used for fencing and garden features or dried out and chopped into logs. Pollarding leaves the more mature willow looking startled by its sudden baldness and not a little unbalanced. It is hard to believe that such a bare bole will ever recover its dignity, until the unstoppable branches shoot up again, thicker and fresher than before. These are the willows in the illustrations by E. H. Shepard of what often tops the polls for the most popular children's book of all time: *The Wind in the Willows*.

The willow's amazing energy means that the springy branches and shoots are generally unfazed by having been rudely snatched from the bole. Thick or thin, willow cuttings are natural survivors, so thrusting them into moist ground will usually result in fresh green sprouts. Within months, a tiny stick can become a sapling, which makes them the easiest of trees to propagate. This is why willows claiming to be descendants of Napoleon's tree could be found all over nineteenth-century Europe, apparently grown from slips of the willows that hung over his grave on St Helena. Today, willow rods are frequently crafted into obelisks and pergolas, arches and wigwams, with the bare brown skeletons quickly disappearing under a thick fur of pale green leaves. An arbour woven from young willow sticks turns from a bare lattice frame to a secluded, verdant bower in a couple of summers.

THE WIND IN THE WILLOWS

Willow poles, used by the Romans to prop up vines, make excellent supports for roses, honeysuckles or even a recalcitrant pyracantha, enabling gardeners to transform the flattest plot into a crowd of flourishing, green, upright figures and forms, thick with leaves throughout the summer to set off the flowering climbers. With careful interweaving through the strong willow twigs, even a climbing rose will go where it is meant to, without the need for wire. Why such vigorous trees should have been associated with broken hearts is therefore something of a puzzle, despite the well-known biblical image. (And this association has now turned out to be entirely misplaced, because the willows by the waters of Babylon memorialised in Psalm 137 are now thought to have been Euphrates poplars.)

Identification of willows has always been challenging because of the proliferation of different kinds and their many shared characteristics. Their bright, lemon-yellow catkins are an early treat for insects, while their powdery seed flies off in all directions with the slightest spring breeze. These are naturally flirtatious trees, cross-pollinating compulsively and giving rise to numerous hybrids. There are some 450 different species across the world – and the botanists are still counting. Even in Britain, the variety is astounding, especially once all the different kinds of sallow and osier are taken into account.

Willows are still grown commercially on the Somerset Levels, where they have always been a major part of the local economy. The best willows for basket-making are the short, stubby varieties, known as osiers, which send up stems of sufficient strength and pliability for bending into wickerwork without a crack or a snap. Osiers come in quite a variety of colours, from Black Maul and Noir de Verlaine to Flanders Red, brown Wissendra or the warm, summery, Golden Willow. Their brightly coloured barks mean that baskets can be woven from a spectrum of natural colours, from peaty browns and purples to oranges, golds and rusty reds. The willow stems can also be dried until they are brown, boiled to the buff and peeled to reveal the light, white wood inside. If baskets make us think most readily of sewing or shopping, the willow's potential as a creative material has recently begun to be properly recognised, with the different varieties providing a renewable palette for natural sculpture.

Black Maul, or *Salix triandra*, was chosen by Serena de la Hey for making a giant willow man in the year of the millennium. Her

SERENA DE LA HEY, *WILLOW MAN*

massive sculpture towered over Somerset like a striding colossus, in celebration of the traditional local industry. The *Willow Man* of the west, caught dramatically with arms outstretched beside the M5 at Bridgwater, is the region's answer to Anthony Gormley's famous *Angel of the North*. *Willow Man* was burned down by arsonists less than a year after his inauguration. Perhaps he had stirred dark memories of the Wicker Man – the legendary druidical practice of binding prisoners inside a huge willow figure and burning them alive in ritual sacrifice? This ancient horror was probably invented by Julius Caesar, who needed justification for his invasion of these islands, and the idea has survived imaginatively through the centuries, receiving powerful treatment in the film *The Wicker Man*. The fire may well have been sparked by a more ordinary destructive response to the sight of a vulnerable giant, or merely the excitement of setting off a beacon that could be seen for miles around. Whatever the arsonist's motivation, *Willow Man* was patiently rebuilt and protected by an enormous moat. Since then he has had only to contend with the local bird population, quick to spy out an easy source of nesting twigs.

The flat wetlands of the West Country are the home of willow sculpture and local festivals are regularly held to promote local growers and to encourage people to try their hands at weaving. The internationally renowned willow sculptress Emma Stothard learned her practical skills on the Somerset Levels and now uses great twisted skeins of willow to create life-size horses, bulls, deer, hares and even a pack of dogs. At Chatsworth House, Laura Ellen Bacon's strange organic forms of woven willow twist in and out of the fruit orchard in the kitchen garden. As the charcoal made from burning willow-wood is the best kind for artists, these sculptures often benefit from this tree before they have left the drawing board.

Julius Caesar was very struck by the versatility of willow and especially its capacity to provide the materials for small, oval boats, which were made watertight by being covered in animal hide. These simple boats were light enough to be slung over a shoulder for carrying across land, making them a prototype of the amphibious craft. The centurions might be trained to form a testudo, or military tortoise, but the beetle-like Britons had tactics of their own. From coracles to carriages, from hot-air balloon cradles to gliders, prams and Bath chairs, basketwork has since been adapted to whatever society deemed a sensible way of getting about.

Willow is also strong enough to make sturdy boxes for fruit and vegetables, linen and logs, bread and bicycles, since different varieties of willow are well suited to the weight of different loads. Wickerwork stools and linen chests, sofas and hanging chairs, shelves and rockers, coffee tables and carpet beaters have all had their moments in the domestic limelight. It is now possible for the ecologically minded to order a willow coffin. With the environmental taxes on plastic carrier bags, the traditional shopping basket (perfectly designed for protecting fragile purchases) may soon be making a comeback. Heavy-duty willow barriers can form a living screen along busy roads to cushion the noise of passing trucks and protect children from the traffic. Lighter hurdles make flexible, perforated shelters within the garden, easy to manoeuvre and less likely to be flattened by strong blasts of wind.

The wood's flammability means that it is now being developed for biomass: in Scandinavia, willow woodchips are already replacing oil as a cleaner fuel for domestic heating and even industrial plants. Willows shoot up at quite a pace, making them suitable for renewable fuels and, as growing willows continue to absorb carbon dioxide, the increase in plantations acts as a counter to the carbon emitted during the burning process. In Britain they are performing important environmental services of a cooler kind. Since osiers can absorb heavy metals as they soak up dirty water, they are often employed in the reclamation of disused industrial sites. With the serious floods of recent years, willows are also coming into their own as defences against sudden inundations, because their long roots, thriving best in moist soil, absorb and clean the water while they stabilise the riverbanks.

Willow-wood has also helped stabilise human beings, since its lightweight and smooth texture made it suitable for fashioning artificial limbs for amputees. Photographs of the Roehampton Hospital during the First World War show large blocks of willow being rapidly crafted to meet the urgent demand for wooden limbs. Despite its combustible qualities, the willow is an ancient remedy for fevers. Willow charcoal is also a friend to those suffering from indigestion and flatulence (the dark grey shapes in a pack of dog biscuits act indirectly as air-fresheners in canine-loving households). The bark of the willow tree has long been regarded as an effective treatment for all kinds of ailments, from sprains to dysentery, with good reason: it contains salicin, the active ingredient in aspirin. The willow's capacity to soothe pain and lower temperatures is a physical property in perfect accord with its gentle appearance. The very sight of a willow, with its delicate leaves beginning to bleach, is enough to calm inner turbulence, dispelling agitation with memories of long, lazy summers and picnics by languidly gliding streams.

For many (and famously Sir John Major), the very name of willow – when hitched to 'the smack of leather' – is synonymous with afternoon tea and Englishness. As everyone knows, the willow furnishes the raw material for cricket bats and has done since the 1780s, when a new variety of white willow, *Salix alba caerulea* was identified in Suffolk. The rich moist soil of East Anglia provides the best conditions for this exceptionally fast-growing variety, which laps up water from the rivers and dykes and provides an especially resilient, wide-grained wood. When the trees are being grown specifically for cricket-bat production, they are carefully groomed to avoid the growth of untidy shoots along the trunk and, after about fifteen years, when the trees have attained a height of sixty feet, they are ready for felling and slicing, first into a series of elongated logs and then into roughly bat-sized blocks. It can take up to a year for the softwood to dry out, but once the moisture has finally evaporated fully, the wood can be compressed to create a bat strong enough to stand up to the world's fastest bowlers. As the cricket willows still grow best in their native county, the batsmen of Australia, Pakistan, India and the West Indies all have to put up with using English bats in Test matches. What is more, the wood of the cricket-bat willow is taken from the female tree.

The willow's natural love of water has enabled it to become a key player in very different arenas. At times of drought, water diviners

would traditionally take a Y-shaped willow twig for dowsing, pacing the earth steadily until the forked rod began to twist and turn downwards, pointing to an underground source. J. K. Rowling was no doubt aware of the willow's traditional importance in witchcraft and the pagan beliefs of Wicca, since she specifies willow-wood for certain wands in the Harry Potter series. Willow magic features memorably in these books, but is most prominent in the writhing form of the Whomping Willow, which is especially active at night under a full moon, distracting attention from Professor Lupin's transformations into a werewolf. Harry and his friends encounter the willow when their flying car is caught in its huge branches, in an episode that reveals another of the more fearful aspects of these seemingly soft, pliable trees. The Whomping Willow's immediate ancestor is the cunning Old Man Willow in *The Lord of the Rings* trilogy, and it is easy to see why these trees have featured as such terrifying characters. A mature willow, with thick, dangling fronds can look like a huge, green tarantula, waiting to grab unsuspecting passers-by and suck them into its cavernous interior. On a moonlit night, the vast, shaggy moving shadow of a tall willow tree can assume quite monstrous proportions.

The moon's eternally changing face and governance of the tides make it a suitable heavenly associate for the ever-adaptable, ever-moving, water-loving willow. The willow's lunar associations run deep, drawing on Celtic and even Sumerian myth. By the time Nicholas Culpeper compiled his classic *Complete Herbal* in 1653, the link seemed incontrovertible: under his entry for the Willow Tree, he wrote, simply, 'The Moon owns it'. Since bruising and boiling the leaves in wine was then recommended for the concoction of a sure-fire antidote to lust, it seems that assumptions about the tree were still being influenced by the classical associations of the moon with Diana, goddess of chastity (this may well have been at the root of all the sad, lovesick songs of disappointed swains). In *Tess of the D'Urbervilles*, the village girls who join the May Dance in their varying shades of white carry wands of stripped willow to signify their virginal status.

The willow's allegiance to the moon made it the tree for the lunatic, the lover and the poet. When Wordsworth looked back on his youth to trace the growth of his poetic mind, the crucial, solitary trip across a moonlit lake began with his theft of a small rowing boat that was tied to 'a Willow tree . . . within a rocky cave'. Heaney, too, remembering all the secret nests he made in the trees at the family farm at Mossbawn, was most attached to the old, hollow willow, whose throat he would climb through into a 'different life'. Once inside, he recalled, 'if you put your forehead to the rough pith you felt the whole lithe and whispering crown of willow moving in the sky above'.

The willow is the tree of inspiration – and this is perhaps the least surprising thing about it. The willow stands next to flowing water, accompanying the sounds of the stream with the natural music of its branches. Often the boundaries between the elements seem to be dissolving, as the dew drops on the willow leaves and the sap from its boughs drips quietly into the waters below. Sitting in a small boat under a weeping willow is like being inside a fountain, and watching the greenery cascade into the darker green of the river. There are no edges and no dividing lines. When the French artist Claude Monet was trying to achieve an ideal of an 'endless whole', unlimited by shores and horizons, he turned again and again to the pond in his garden at Giverny. It was as if he was painting with plants, as he filled the pool with water lilies and surrounded it with weeping willows. When he was not planning and planting, he was painting his pool, creating images in which the planes and surfaces dissolve, as the multi-petalled flowers float on reflections of the trees, and the fronds of the willows make waves more visible than the water. This tree adapts so rapidly that it can be taken for water, sky or earth. With shifts so swift, the willow's whispers can be heard again and again, but even the most attentive listener will never quite catch all that they may be saying.

HAWTHORN

Feathery leaves and clouds of blossom give the hawthorn a soft appearance, but hidden beneath the inviting green and white are inch-long, flesh-shredding spikes. Not many gloves are completely thorn-proof and even the saplings bristle with little swords. It is not that these trees are on the attack – they are just looking after their own. In the deadest months of the year, when other gaunt, bone-picked trees are raked by freezing blasts, the hawthorn's tangling, criss-cross branches still offer a haven to half-starved birds and wildlife. A little later on, the hawthorn – or May, as it is also known – makes the perfect site for songbirds to roost; for though their irrepressible voices cannot help but attract attention, the surrounding ranks of spiky palisades will fend off all but the most determined predators from the birds' nests. John Clare loved watching thrushes patting their clay-lined cradles 'within a thick and spreading hawthorn bush' because he knew it was the safest place for the shining pool of sky-blue eggs. The hawthorn offers natural protection to whatever is hidden within, but its peculiar defensiveness tends to trigger retaliation. Centuries of determined human intervention have combined to make it doubtful whether this can really be a *tree* at all.

If left to its own devices, a solitary hawthorn will grow into an elegant, round-crowned, slim-limbed tree of thirty feet or more, or it might remain squat like an arboreal hedgehog, with quills brushed and coloured according to the season. In exposed spots, hawthorns can balance at dramatic angles, crowns streaming out in striking

HAWTHORN, FROM HUNTER'S EDITION OF EVELYN'S *SYLVA*, 1776

silhouettes from bracing themselves against the wind over many years. None of this natural beauty is of much interest to farmers, however, who see in the tree's natural exuberance and dense prickly character the ideal materials for a hedge.

At the time when timber offered the most lucrative return on land, the hawthorn was hardly regarded as a tree in its own right, but planted only as a lowly security guard for taller, more valuable species. (The very name comes from *haga*, the Old English for haw, which shares Germanic roots with *hecg / hegge* – hedge.) In Evelyn's *Sylva*, the hawthorn – also known as the 'Quick-set' – is introduced briefly in a chapter on fences, where the main preoccupation is with the practical business of setting the saplings as swiftly as possible. Later editions of the book do display a beautiful, full-page engraving of a sprig of May blossom, as if to query Evelyn's dismissive tone, but as far as he was concerned the hawthorn's primary purpose was to protect timber trees. Oak, in particular, is vulnerable when young, because cattle, deer and rabbits all find the fresh, crispy leaves of an oakling irresistible. The answer to such persistent incursions was to surround timber plantations with an impregnable blockade of hawthorn.

A line of hawthorn saplings does not look like much of a deterrent, admittedly: these trees are rather ungainly in their youth, often resembling a fistful of spindly bottle brushes more than a proper hedge in the making. They do fling themselves out boldly enough, reaching wide like Vitruvian Man, but there will always be a stray, straggling limb, not quite in perfect proportion. What these gangling youths need to be transformed into a strong, well-disciplined line of defence are skilled hedgers, who hack and plash until they are laid low. It is only after such brutal rites of passage that the hawthorns turn into a passable, or rather impassable, hedge.

Every year, the National Hedgelaying Championships offer a chance for hedgers to gather from all over the country and show off their skills, competing in regional classes in the hope of becoming the supreme champion. Within hours, unlikely stretches of unkempt trees are transformed into the neatest living fences imaginable. It is not a quaint or quiet pastime. Modern hedgelayers set forth with all the determination of an ancient hero going to do battle with the Gorgon, but even the most fearsome, myriad-headed green monster succumbs fairly quickly to a chainsaw. It is, of course, important to remember that the competitors are *laying* rather than *slaying* the hawthorns – by slicing at the trunk close to the ground, the tree can be pushed down at an angle but not completely felled. As each hawthorn falls down across its nearest neighbour, so the solid barrier of diagonal 'pleachers' is formed. It might sound easier simply to chop off the hawthorn crowns, but this would defeat the main purpose of a hedge: decapitation leaves gaps all along the bottom, producing something more like a dental X-ray, full of low tunnels for sheep and rabbits to wander through.

Different hedging styles have developed across the British Isles in line with regional materials, farming practices and the lie of the land. In Devon and Dorset, hedges are generally laid on top of banks to keep sheep safe, but across the Midlands they have to be sturdy enough to avoid being inadvertently demolished by a herd of cattle. This means using the 'Midland bullock' method, an elaborate system of stakes that are driven in beside the hawthorn stumps and then woven together with hazel binders for maximum support. In the Welsh Borders, the stakes are angled instead and the living hawthorn stems layered with dead brushwood to deter hungry sheep from munching on the fresh shoots. It is all rather labour-intensive, but time spent plashing is a sensible investment: a well-laid hedge will last for fifty years, with only minimal annual trimming to keep it in good order.

These field barriers are often signs of how well neighbouring farms are run. A boundary hedge will quickly begin to resemble a Mohican when one side is clipped smooth and the other just left as a tousled bed-head. Hedges can be thick and orderly or more like lines of sportive wood run wild. On the horizon, a hawthorn hedge can resemble a great, unbroken, well-lagged pipeline or a row of rather messy, green inkblots. Modern farmers have more instant options for dividing their land, but neither barbed wire nor electric fences prevent the soil from washing away or provide shelter to sheep in winter and shade for cattle on hot summer days. Unlike wire, hawthorn is also a good source of firewood. The hard logs of the hawthorn create the hottest fires, as well as giving off an eerie lilac glow, while the brushwood left over after hedgelaying makes excellent kindling. Since a hedgerow is home to countless birds, insects and small mammals, the farmer who sets a hedge is also the founder of a whole community, even though he is in the habit of sending waves of fear through his subjects on a fairly regular basis.

As hawthorn wood is hard and plentiful, it was traditionally used for hammers or walking sticks, the handles of daggers and the teeth of mill wheels. Even the roots were crafted into combs. The thorns provided natural hooks, fishing tackle, needles, pins, winkle-pickers and even a makeshift stylus for vinyl records. A spiky branch will prevent cats and pigeons from turning a flat windowsill into a landing site. The beautiful, fine-grained, rosy-coloured wood of the hawthorn can be polished into glimmering boxes or candlesticks, as well as furnishing decorative veneers. The signature haws, or scarlet berries, also have practical benefits, as they can be pulped and strained into a sharp jelly or soaked in brandy to create a powerful liqueur. Taken in the form of a tonic, they will combat coronary problems and arrhythmia by increasing the flow of blood to the heart and perhaps by lowering cholesterol levels. There is some evidence to suggest that the haws may also be rallied against dysentery or tapeworm.

Hawthorn trees have been recruited for service in their thousands. No other tree has done more to change the entire face of Britain. 'The English countryside' means rolling hills, golden church steeples and a patchwork of colourful fields marked out by hedgerows and trees. Whether driving along a country lane in Herefordshire, flying low over the Yorkshire Wolds or taking a satellite view of the Cotswolds, the quilted cover is instantly, reassuringly recognisable and rural. But what strikes us, now, as the natural landscape of Britain is really a palimpsest of older farming practices. From the earliest forest clearances to the most recent housing estate, the land has been shaped to human needs, and so beneath the familiar scenery is a less visible history of shifting land ownership.

During the Middle Ages, villages were surrounded by large open fields, divided into strips for local people to farm, but as land changed hands and agricultural methods improved, Britain underwent a wholesale restructure. The higgledy-piggledy patterns of the past had no place in the geometric efficiency of the highly productive, modern estate and, since new fields required clearly defined

THE
LEGACY
OF
ENGLAND

THE LEGACY
OF ENGLAND

BATSFORD

BRIAN COOK BOOK JACKET, 1935

boundaries, the age of enclosure was also the age of hawthorn. The later eighteenth century saw mile after mile of hawthorns being planted in long lines, straight-edged in regimental squares and rectangles, enclosing millions of acres of open land. In little more than a century, rural Britain took on the character that would come to be lovingly portrayed in countless paintings, woodcuts and photographs, gradually turning from the new into the nostalgic. *Shell Guides*, railway posters and Batsford books would all affirm that, except for the upland regions, Britain was, and always has been, a country of hedgerows.

Hawthorn still marks out the familiar features of much of the country, which is why uprooting hedges can be such a shock for local residents. It is much more unsettling than meeting an old friend who has shaved off his beard; more like the sense of loss that can come with chemotherapy – a sudden exposure of unprecedented vulnerability. In the 1960s and '70s, the shocks came thick and fast, as the increase in crop yields and mechanisation meant that, in the most productive arable regions, many hedges were torn up to create fields enormous enough for modern ploughs and combine harvesters. With the consequent grain mountains and EEC agreements, whole fields were subsequently set aside from intensive agriculture, their hedges left to become frayed. Far from being an unchanging feature of the British landscape, the hawthorn has had a history as chequered as the landscape it defines.

Although unproductive fields are dismaying to some, I must confess to a personal soft spot for all those hawthorns that have finally succeeded in bursting free from years of hedgedom and are now reasserting their identities as trees. You can almost feel the sense of release as the branches fly in all directions. Neglected hedges running parallel along a disused track will stretch out and touch boughs, vaulting into arches. Once they are feathered with spring leaves, the entire track is transformed into a tunnel of green light, a secret world, protected and quiet, where a young deer might stop and stare before disappearing through a wall that is no longer impenetrable. The old hawthorns are still bent and disfigured from their years of submission, but their spreading bundles of trunks bear young, slim branches, undistorted by human management. Light falls through these wilder forms onto patches of fresh grass or a rusting harrow abandoned years ago. Where the forgotten path meets the well-trained, hedge-lined road, there might be a trail of honeysuckle or an empty yogurt pot, a clump of cow parsley or a polystyrene chip tray.

The hawthorn harbours all sorts of unlikely things. Old tales about pots of gold under thorn trees are legion, probably because thieves were less likely to stumble on secret subterranean store-houses with this natural guard standing by. The name of Hawthorn Hill near Bracknell, for example, is said to commemorate the lucky discovery made by a local resident. After dreaming that he would make his fortune in London, he set off for the city in search of trea-sure. When he arrived, he met a stranger who told him of his own dream – about finding a pot of gold under a hawthorn tree, before laughing off the very idea of taking dreams seriously. The Bracknell man went home again and decided to dig beneath the thorn tree on the hill, where, of course, he finally found his treasure.

Mystery hangs about the May tree, but it was at the Battle of Bosworth Field in 1485 that it yielded the biggest treasure of all, when the crown of England was discovered suspended from the hooked branch of a hawthorn. Quite how it found its way there – or, indeed, if it really did – remains open to debate, because although the discovery of Richard III's crown and the immediate coronation of Henry Tudor on the battlefield are recorded in contemporary chronicles, the tree is not. The victorious Henry VII was so quick to incorporate the hawthorn into his new heraldic badge, which depicted the crown above a thorn tree, that the legend instantly took root, as hawthorns do. As the battle took place on 22 August, it is possible that the local thorns would have been covered in bright haws at the time, especially since the introduction of the Gregorian calendar has shifted our sense of the seasons a little.

Irrespective of the weather in 1485, a stylised hawthorn made a memorable symbol of victory. The hawthorn, with its spiked, blood-spattered branches is the tree of battlefields. For Henry, whose claim to the throne was not as strong as he might have wished, the hawthorn was certainly intended to encourage openness to a new order of things. Since it combines the colours of the warring houses of York and Lancaster in the course of its annual cycle of white blossom and red berries, this hardy native tree was providing sound stock for grafting on the dynasty that would succeed them both. The design of the crown above the thorn drew on the powerful symbolism of the Crown of Thorns, underlining the divinity of the monarch and the promise of a new, better reign. The hawthorn's prickly reputation also reminded the people of Britain that this new king was ready to defend his claim against all comers.

The association of hawthorns with the passion of Christ has been reinforced by the tree's striking natural habits: the sudden profusion of white spring blossom. Every year, huge heaps of flour seem to be dropped along the branches by a supremely careless cook. Almost overnight, the hawthorn turns from tufty spring green to thick white. The May tree proclaims the arrival of spring, wrapping itself in snow as if to mock winter into retreat. In the gloom of 1943, Stanley Spencer expressed his faith in the annual resurrection of life in his painting of *Marsh Meadows*, where three hawthorns, glowing and white, light up a field at Cookham. This was the tree known in some areas as the 'Awe Thorn', and not just on account of the local accent.

For David Hockney capturing the annual changes of the Yorkshire landscape, the sudden blossoming of the hawthorn means 'action week'. At his landmark exhibition at the Royal Academy in 2012, a whole room was filled with paintings of huge, disturbing, custard-coated trees. In *May Blossom on the Roman Road*, the hawthorn forms are magnificent and monstrous, their creamy sprays like enormous caterpillars or even maggots crawling all over a great green corpse. And yet, the profusion of unexpected colours conveys the sudden excitement of the moment when everything springs into life, banishing the understatements of winter. Hockney's celebratory paintings magnify the roadside presences and release the ancient power of the hawthorn, rendering this common tree a thousand times lovelier, more dangerous than anyone would have guessed.

Part of the hawthorn's mysterious power lies in its unpredictability. In the unkind spring of 2013, after what seemed like an interminable winter, it was early June before the hawthorn deigned to brighten the fields of rural Buckinghamshire. The natural calendar is entirely weather dependent, so May blossom might come hurrying out in Penzance by April or fail to arrive in Aberdeen until midsummer morning. Whenever it appears, the blossom is all the more arresting for its unpredictability. That the fickleness of the hawthorn has always caused mayhem can clearly be seen when Theseus and Hippolita go out into the forest to perform the rites of May in *A Midsummer Night's Dream*.

The sheer surprise of 'May' blossom sheds light on the veneration inspired by the old Glastonbury thorn, which flowers every Christmas and again in the spring. According to the legend, Joseph of Arimathea left Jerusalem after the crucifixion and travelled all the way to Britain, ending up in the West Country. As he stuck his staff down on Weary-All Hill at Glastonbury, it burst into a thorn tree. For centuries, the sacred thorn continued to flower in the nativity season and again in Holy Week, like a reliable miracle tuned to the patterns of the abbey church. During the Civil War, a soldier loyal to the new, puritanical ideals of Cromwell, and shocked by anything that smacked of superstitious idolatry, took an axe to the Glastonbury Thorn. As the stump still furnished a few twigs, a cutting was quietly taken and replanted, so in due course the hawthorn grew again, apparently demonstrating a capacity for eternal life. Every Christmas, a sprig of the flowering thorn is sent to the Queen – or was until 2010, when the ancient thorn was decapitated again, this time with a chainsaw. Since then, new thorns, nurtured from the original tree, have been planted again and subsequently vandalised. The battle over the Glastonbury Thorn is becoming a regular news item, which may in turn be generating its own succession of episodes.

THE GLASTONBURY THORN

The old abbey is a Christian site, and the ancient thorn's habit of coming into bloom to celebrate Christ's birth and resurrection fits perfectly with the ecclesiastical year. Its miraculous character is explained by the particular species of hawthorn – *Crataegus monogyna* 'Biflora', which has a winter and spring flowering season, perhaps originating from early grafting. Glastonbury is also a sacred site for those with pagan beliefs, associated with natural religion and the ancient seasonal cycles. The sudden delight of blossom and fragrance in the month when daylight is in short supply naturally inspires awe in those witnessing this unexpected renewal. Repeated attacks on the thorn cause deep distress to members of both the established Church and pagan religions, but whether the tree is really a victim of more militant tendencies on either side, as is sometimes suggested in the press, is unclear. The impulse to destroy any ancient site or beautiful natural phenomenon is hard to fathom, and it may be that the poor flowering thorn suffers from its own excess of meaning – stirring vague but deeply felt prejudices, jealousies or fears which erupt into aggression.

The powerful emotions still stirred by the Glastonbury Thorn are redolent of older, unwritten lore. To bring hawthorn blossom into the house was – and, in many minds, still is – regarded as foolhardy in the extreme. Despite its beauty, the flowering thorn brought bad luck and even death on the family. Hawthorn blossom smelled of rotting corpses and the great plague (the distinctive smell has now been identified as trimethylamine, a chemical also produced by decaying human bodies), so it is not surprising that this was not a popular choice for flower arrangements. Richard Mabey has suggested that the white blossom of this tree smells of sex, which may be another reason why not everyone wanted it hanging heavy in the air. There is something more to the pervasive fear of May flowers than a troubling smell, however. The urge to keep the hawthorn out of the house may not be so different from the hedgelayer's urge to keep it down.

Thorn trees can live for centuries, gathering stories and superstitions with the passing years. In the churchyard at Hethel, near Wymondham in Norfolk, one of the oldest hawthorns in the country still stands as it has done since the thirteenth century. According to James Grigor, who visited the parish in the early nineteenth century, and also Vaughan Cornish, who undertook a later study of the nation's oldest thorn trees, this ancient thorn, with branches so old and hollow that a man's arm could be plunged right inside, is known as the 'Witch of Hethel'. However, this strange Norfolk thorn has also been claimed as another living heirloom from Joseph of Arimathea. Local lore, though often retaining truths of its own, is not always the most reliable source of historical fact. The recurrent story of Joseph's staff does reveal the way in which the Church encouraged the assimilation of earlier narratives. Stories of holy thorns probably signal the sites that meant most to pre-Christian culture and which continue to attract those whose spirituality is rooted in the natural world.

In the Irish ballad of 'The Fairy Thorn', the old hawthorn tree exudes a chilling power. The gnarled old thorn enchants the girls who dance around it, so that the fairies – a much stranger and more threatening race of beings than Cicely Mary Barker's pretty 'flower fairies' – can snatch one of them away. Respect for the fairies means that golf clubs in Ireland, such as the one at Ormeau, always leave veteran thorns in place. Road builders in Ireland are very wary of cutting down these dangerous trees. The safer course is to route the road around the thorn, which is the reason for the slightly odd positioning of one of the slip roads from the motorway between Antrim and Ballymena. The most celebrated hawthorn in Ireland is in the west, near Ennis, and it marks the traditional rallying point of the Munster Fairies. The new motorway between Galway and Limerick was delayed by almost a decade as local planners wrangled over how to avoid uprooting the ancient thorn. Eventually a new route was agreed, but so was the caveat that no traffic should pass within five miles of the thorn, which now commands its own protected area.

Traditional suspicions of these trees animate Wordsworth's unsettling poem 'The Thorn'. An old hawthorn, little more than a 'mass of knotted joints', is introduced as a 'wretched thing forlorn', but, far from inspiring pity, it exudes an aura decidedly sinister. Inspiration for the poem came from a stunted hawthorn on a ridge in the Quantocks, but in Wordsworth's ballad, the unremarkable feature is charged with compelling mystery. The garrulous narrator of the ballad points an accusatory finger at Martha Ray, the woman in the red cloak who sits alone by the thorn, crying 'Oh misery! oh misery!' The poem, however, conveys a sense of more powerful forces at work, fixing attention on the fantastic colours of the mossy heap beneath the thorn:

> All colours that were ever seen,
> And mossy network too is there,
> As if by hand of lady fair
> The work had woven been,

And cups, the darlings of the eye,
So deep is their vermilion dye.

Ah me! what lovely tints are there!
Of olive-green and scarlet bright,
In spikes, in branches, and in stars,
Green, red, and pearly white.
This heap of earth o'ergrown with moss,
Which close beside the thorn you see,
So fresh in all its beauteous dyes,
Is like an infant's grave in size
As like as like can be.

The strange mound, just the size of an infant's grave, glows with the colours of a flourishing hawthorn, even though the stunted tree above is old, grey and leafless. Though the superstitious narrator is preoccupied with Martha's suspected murder of her child, Wordsworth charges his poem with primordial fears of the hawthorn's terrifying powers. The treasure buried beneath his thorn is such that no one would want to find.

The hawthorn was the tree of the white goddess of ancient native mythology: enticing, entrancing, but potentially deadly. The milk-white flowers of the May tree, so soft, so inviting and yet so rebarbative, have maddened mankind. For all its associations with blood, battles and barricades, the real fear surrounding the hawthorn is of some terrifying feminine force that lurks within this formidable tree – or, rather, of the minds of those who feel so threatened by its natural beauty. No amount of plashing and binding can quite beat the May tree into submission.

The old Celtic festival of Beltane took place in early May to mark the change of season and welcome the sun. May bushes would be hung with bright shells, ribbons and flowers as an offering to the Sidhe, the capricious, faintly malevolent, supernatural beings who passed through fairy mounds between their world and this. Whether the tree was thought to offer protection or invite propitiation is not clear, but a sense of ill omen tended to hang about the thorny boughs. The proverbial warning, 'Marry in May, repent for aye', is surprisingly widespread, given the voluptuous loveliness of the month and its tree. It is telling, too, that the international distress signal, though derived from the French request for immediate help, *m'aidez,* was so rapidly anglicised into 'mayday, mayday'. Beltane rituals lingered long, feeding ideas about the May tree and its annual festival, though the very innocence of the celebrations suggests concerted efforts to regulate the passions of the season. The first of May is ushered in with early-morning singing, floral parades and May breakfasts to turn the event into a wholesome family affair. Children are taught to create elaborate multicoloured lattices by dancing in and out, backwards and forwards with long, bright ribbons hooked onto a tall maypole, until it begins to resemble a giant cupcake. The hawthorn is central to the celebrations, adorning May queens and May houses, but the tall poles, with their strangely scented white blossom, still hint at wilder energies.

The residual power of the May tree continues to be felt and even urban modernity still retains the traces of an older, not quite vanished world. There was a time when people arranged meetings according to the location of the oldest hawthorn in the parish and these ancient marker trees have left a permanent legacy. Dozens of London streets have names deriving from vanished hawthorns – Thorn Street, Thornton Road, Thornhill. At Bristol University students and staff congregate every day in the bar of the 'Hawthorns', though there is no longer any thorn tree to be seen. Fans of West Bromwich Albion, too, regularly head for the 'Hawthorns', because their home ground was built on the site of a hawthorn field. The club's distinctive badge depicts a song-thrush perching on a clump of hawthorn leaves. When Langley Bush, the famous old hawthorn near Helpston, was destroyed, John Clare wrote wistfully of the tales told of it by the gipsies and shepherds, but he also knew that it would be 'long ere its memory is forgotten'. Even when an old thorn has been completely rooted up, it seems, something still lingers in its place.

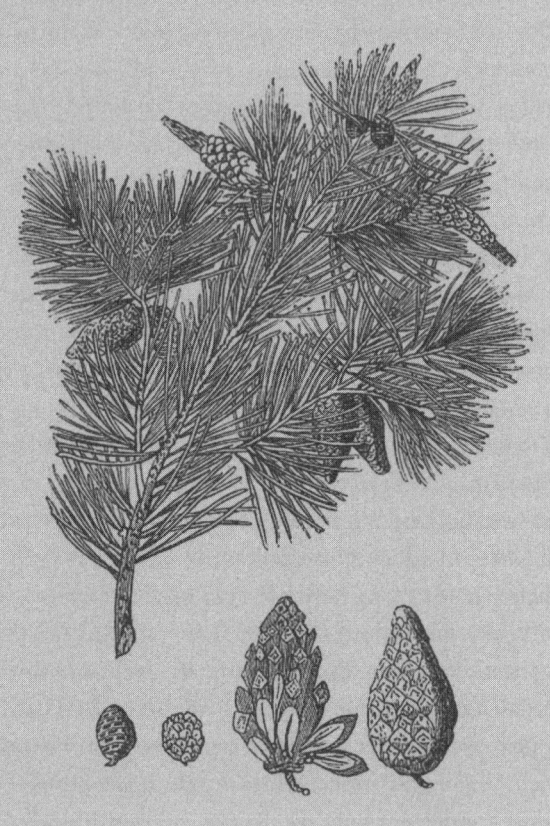

PINE

❋

OUR chest of drawers was called 'Violet'. Most items of our family furniture had names, and this one was inspired by the thick, purple gloss paint that caught my mother's eye when she was scouring a local junk shop for hidden treasure. In the long hot summers of the mid-1970s, surfaces began to melt away and people started to strip. My mother was not part of the new wave of streakers, racing through parks or leaping over cricket wickets, but she was a pioneer of the new fashion in furniture. We were all surprised to see Violet on her return from a dip in an acid bath – clean, bare and golden-brown in her newly revealed wood-grained beauty. It was time for a new name.

As a new ideal of living closer to nature began to take hold, so did the demand for stripped pine. Self-sufficiency meant paring away unnecessary layers of urban living to recover an essential naturalness. Tables, dressers, shelves, plate racks, bedsteads, wardrobes, lamp-stands, towel rails and even toothbrush holders – there was apparently very little that could not be made from pine. The return to natural materials in the 1970s signalled a reaction against the polythene, plastic, polyester space age. If modernity meant man-made materials and high rise, then it was definitely time for pine. Even the most cramped inner-city flat could still have its own country kitchen. Instead of linoleum or Formica mimicking wood, everyone wanted the real thing and so the old washing dollies and roundheaded pegs that the previous generation had been so glad to replace were now revived as decorative features. New furniture stores sprang up to fulfil the dreams of bare, semi-Scandinavian style.

A decade or two later, when some of these shiny piney pieces began to look as if someone had been a little over-generous with the fake tan, it was time to opt for more sophisticated waxing – or to cover up again. Those who returned their stripped shelves to tasteful shades of chalk white or duck-egg were just as indebted to pine, though, since both the oil in emulsion paint and the turpentine needed for cleaning gloss-clogged brushes after a hard day's DIY are extracts of this astonishingly versatile tree.

Pine trees are one of Britain's oldest native arboreal species. As the glaciers retreated after the Ice Age, some ten thousand years ago, the pine spread north through England and Wales to take firm root in Scotland, as it did from Scandinavia, across the far northern hemisphere as far as Siberia. There are over a hundred different kinds of pine, but the most ancient British settler was *Pinus sylvestris*, the aptly named Scots pine, which thrives in the rocky terrain and thin, acidic soil of the Scottish Highlands. The tree's striking beauty also made it a highly desirable feature for any self-respecting Victorian estate, which suddenly began to need its very own pinetum. Since the pine's generous cones seed so easily, wild saplings quickly sprang up in surrounding areas, making the tree familiar to people right across Britain. You can often spot a group of Scots pines in an English municipal park or an old rectory garden, sticking up above their shorter, bosky companions like a chimneysweep's brushes, as if slightly startled at reaching the open air. Sometimes you come across them beside a country road, their peachy-patchy trunks stretching on high, like a gaggle of vast flamingos, dressed up in fur capes.

The best place by far to see Scots pines is in their native land – towering tall and strong over an empty loch, with their branches high above the rocks, held out straight to carry dark, tufty crowns. The pine is a commanding presence, whether standing in solitary grandeur on a bare precipice in the Highlands or dwarfing human beings in a stately wooded valley. On a still day in the Galloway Forest, lochside pines appear to stretch so far into the surface of the water that in the reflections the trees seem taller than the hills

AGNES MILLER PARKER, RED SQUIRREL AMONG PINE CONES

beyond. The pine was an obvious candidate for Scotland's national tree, and during the closing months of 2013, with devolutionary energies gathering momentum, the Forestry Commission for Scotland ran an extensive consultation on the nation's arboreal preferences. The Scots pine garnered 52 per cent of the vote, placing it head and shoulders above its nearest rivals, the rowan and the holly – a position to which it is not unaccustomed, of course. Patriotic fervour also gave the Lady's Tree, an elegant Scots pine at Dunkeld, famous for providing a safe nesting ground for ospreys, the edge over other noble competitors in the Scottish Tree of the Year competition in 2014.

Pine trees have evidently been integral to other parts of Britain for a very long time nevertheless. The massive storms that shook the newborn months of 2014 left the coast of Britain looking battered and unfamiliar. Among the most startling transformations was the reappearance of the ancient, submerged forest near Borth, in Cardigan Bay. When the huge tide began to recede, a stretch of beach was left, filled with what looked first like fins – strange, dark, angular shapes, emerging through the water. Gradually they seemed more like a vast army of spectral horses and helmets rising slowly from the mud, as if waking from centuries of sleep. They were, in fact, the remnants of the prehistoric forest that once covered the western side of what is now modern Wales. These were the stumps of ancient oak and pine trees, preserved in the peat for thousands of years. The old story of Caintref Gwaelod, the submerged forest, drowned through the negligence of the ancient Welsh royal family, was suddenly turning from myth into history.

Although it is rare for the old legends to take quite such concrete form in 21st-century Britain, forests are deeply rooted in our imaginations, often feeding on unfathomed fears. Vast pine forests feature in many fairy tales, usually as places of unspecified menace – easy to enter but seemingly impossible to escape. Wariness of the dark wood is instilled in young children as they listen to the tales of Hansel and Gretel, Little Red Riding Hood or Beauty and the Beast. Even though the story usually ends happily, it leaves a dim sense of some terrible menace lurking within the evergreens. Much of northern Europe is thick with ancient legends of the Mirkwood, the Ironwood and the *Schwarzwald*, or Black Forest. The pine forests of Russia are home to both Baba Yaga and the wolf that devours Peter's duck in Prokofiev's fable. Witches, wizards and wolves all work their powers

LITTLE RED RIDING HOOD

in dark woodlands, but pines offer particular peril, because under their black canopies the trunks stand invitingly apart with promises of almost tangible sunlit clearings and cabins.

Forests are the natural habitat of bears and wolves: from the temperate mountain forests of Alberta to the Russian taiga, brown bears live, hunt and rear their cubs, using the furrowed trunks of tall pine trees as scratching posts. Wolves thrive naturally in many different habitats, but since their main predator is man, they have retreated into those with the best hiding places – remote wooded mountain ranges. With the steady eradication of wolves even in the deepest forests, the creatures that once struck terror into the hearts of local farming families became objects of pity rather than fear and are now a protected species in the few European countries where they survive. (Campaigns to reintroduce wolves or bears into their ancient territories still meet with opposition nevertheless, as unforgotten anxieties rear up.) The last wolf in Britain was reputedly roaming the pinewoods of Scotland until it was finally killed in the eighteenth century, transformed at a stroke from terrifying predator into mythic beast.

Once Scots pines covered large tracts of the Highlands, though the legendary Caledonian forest may have been less extensive than has sometimes been imagined. There are still a few surviving pockets of ancient pine, in Glen Affric, on Loch Rannoch, in Speyside or Loch Maree. These remnants of wildwood, little changed since the aftermath of the Ice Age, offer sanctuary to some of the rarest natives, such as the red squirrel, pine marten and Scottish crossbill. Although many are drawn to the idea of a place completely untouched by humankind, the very act of imagining it assumes the presence of at least one human observer. This sort of 'return to nature' is an unrealisable fantasy of an existence that never was, though its enduring power is obvious enough. Utter isolation cannot exist in this densely populated archipelago, but modern, sociable, predominantly urban dwellers can come as close as possible to the primordial character of prehistoric Britain by dreaming of the old pine forests of the north.

The pine has inspired deep admiration in those lucky enough to contemplate its natural magnificence. John Evelyn, the great seventeenth-century tree watcher, was so filled with wonder at the remoteness of some of the old pines in Scotland that he thought they must have been planted by God and filled with 'blessings' that were yet to be understood. When John Muir, the pioneering conservationist, first travelled alone in the mountains of California, he was deeply moved by his first encounters with the majestic sugar pine (*Pinus lambertiana*), 'silent and thoughtful in sunshine, or wide-awake waving in storm winds with every needle quivering'. These silver-topped giants, reaching heights of 250 feet, and their rugged companions, the yellow pines, struck Muir as 'the very gods of the plant kingdom'. Muir was born in Scotland, but after emigrating to America as a child, he devoted his entire life to exploring and celebrating the extraordinary natural phenomena of his adopted country. He found the wide expanses of America over-flowing with God's presence and, among the vast, forest-covered steeps of the Sierra, he listened to 'the preaching of pine trees' and wrote down the 'sermons on the mountains'. Those who had been living in California for centuries, the Achomawi, were equally rever-ential towards the sugar pine, whose seeds represented the origin of mankind, dropped from the hand of the creator of all things.

When Muir first visited in the 1860s, Yosemite was an uncon-taminated wilderness, but by the end of the century he was lamenting the destructive forces of sheep-farming, legislation and the lumber industry and campaigning for the creation of National Parks to conserve the natural beauty of the region. Muir's awe in the presence of the vast sugar pines was deepened by the growing aware-ness of what had happened along the North Atlantic coast of America, where, in a matter of decades, the vast forests of white pine (*Pinus strobus*) had virtually disappeared. These trees were sacred to the Iroquois and represented the most distinctive of the indigenous phenomena. Yet, they fell before the inexorable axes and sawmills of the nineteenth century, casualties of international naval conflicts and the railway age.

Pine timber was an extremely valuable commodity, and by the time Muir was experiencing the sublimity of the mountain forests, Californian pines from the Monterey peninsula had already been shipped off to New Zealand for planting in the new British colony. The glorious, shimmering sugar pines of the Sierra Nevada were

YOSEMITE NATIONAL PARK

majestic, but vulnerable. Pine trees were at once the living emblems of an unspoiled Eden and the objects that most attracted commercial interest. This tree was bringing about the destruction of paradise simply by being there. The National Park campaign eventually secured Yosemite from commercial logging companies, but even after being designated a nature reserve its famous beauty attracted thousands of visitors eager to see for themselves a landscape untouched by human hand.

The surviving patches of wildwood in Scotland give rise to smaller-scale, but oddly similar, ironies. Though now protected from timber merchants, these ancient Highland pine stands attract attention of a different kind. As more admiring visitors arrive, the chance of damaging the fragile habitat and frightening its shy, scarce creatures into deeper retreat is ever increasing. Raised awareness of ancient woodlands may contribute to their survival, but it may also act as a spur to tour companies and developers to expand their interests into more inaccessible places. On the other hand, neglecting the significance of ancient pinewoods may be even more destructive, as Muir saw when the sheep farms began to encroach into California's wilderness.

Conservationists are faced with daily dilemmas. Although leaving woodlands alone can have many ecological benefits, because fallen trees are home to insect colonies and fungi, as well as providing the nutrients for regrowth, problems often arise when some species are allowed to roam freely. When deer, for example, live in wooded areas without natural predators, the danger is that they will destroy too much of the vegetation and then starve. Responsible forest management is a major contemporary challenge – and, of course, as soon as things are being managed, the very idea of ancient wildwood starts to seem a little contrived. People have been working woodlands for a very long time and the pine's useful versatility means that pinewoods have been especially subject to human intervention of one sort or another. Wildwood is, in some ways, the least natural kind of pine environment, even if it is the most appealingly unspoiled.

Indeed, the relationship between people and pine trees is so intimate and ancient that it is difficult to regard it as anything other than 'natural'. The oldest living pines in the world, the bare, twisted Californian bristlecones (*Pinus longaeva*), have been alive for some

five thousand years, and the best known has been named affection-
ately as Methuselah. In the celebrated Chauvet cave in the Ardèche,
the world's oldest murals show that Neolithic artists, working in the
eighteenth century BC, were using charcoal from pine trees that
must have been living some 32,000 years ago. For twenty millennia
at least, it seems, man has been chopping up and cultivating this
tree. Such a long-standing symbiotic relationship does raise the
question of just what the 'natural' state might be.

For the paradox of the pine is this: the tallest, most stately, elegant
and mysterious of all trees is also the one routinely employed for the
most menial tasks by humankind. It looks like a thoroughbred, but
it is the workhorse of the woods. From Celtic folklore through to
modern forestry, the pine has always been admired not so much for
its beauty as for its utility. Pine is the ultimate multitasker – those
long, straight trunks have in their time provided masts for tall ships,
pit-props for mine-shafts, telegraph poles, fence posts, rafters and
railway sleepers. Pine saplings rapidly grow into tall, strong towering
trees, ready for felling, stacking and shipping off.

In many regions across the globe, pine is the most readily available building material and seems almost designed for human
constructions. I once stayed in a log cabin in the Highlands, which
felt a bit like living in a packing crate because the floorboards
matched the roof and the wooden walls. Lying in bed was like
floating on the ceiling, looking down on the floor, especially after a
glass or two to keep out the cold. This was another of those attempts
to be close to nature, which led as readily to thoughts of worldwide
industry. A later trip towards the Arctic Circle, along the ice-packed,
underpopulated roads of Norway, left memories of head-on encounters with massive lumber trucks just as permanent as memories of
the Northern Lights.

Many of the world's great rivers have in their time been turned
into thoroughfares for huge floating rafts of logs. Cities in Wisconsin
or along the Mississippi grew up around the riverside timber yards
and the paper mills, which were originally powered by the flowing

LOGGING IN WISCONSIN, 1885

water. Paper mills depend on pine for their key ingredient, wood pulp, because pine timber is soft and shreddable and relatively cheap. Papermakers discovered many years ago that sizing paper with rosin, the solid substance created from heating the resin from a pine tree, helped to ensure maximum smoothness and minimum absorbency. Extracting the resin from pine trees also meant that when the wood was burned, the soot was much drier and better suited to manufacturing ink.

Rosin, or colophon as it was once known, because the finest kind came from the pines of Colophony on the Aegean, is still in demand in the world of classical music. When rubbed across the bow of a stringed instrument, rosin reduces its tendency to slip on the strings; when applied to ballet shoes, it lessens the chances of embarrassing mishaps. The gleam on a violin also results from a pine-derived varnish, which means that anyone listening to Sibelius' Violin Concerto and feeling as though they are being wafted into the pine forests of Finland is quite right – the music is more closely connected to the pine trees than might immediately be apparent. Rosin can also be used to glaze chewing gum – not that anyone would resort to gum at a classical concert.

Paper-sizing with rosin, as a Victorian innovation, is a relatively recent development in the long history of human relations with pine. Almost as soon as people began to build boats, they started plastering them in pitch to prevent the water seeping inside. The reason British sailors were known as 'Jack Tar' was because the ropes and rigging of naval vessels were regularly treated with tar extracted from pine trees. Pinewood is dripping with resin, so the slow burning of the logs and roots that turns trees to charcoal also produces tar and pitch – those thick, black, pungent, viscose substances, which begin to ooze out as the sap is distilled. Malleable enough when warm, tar and pitch will stick to surfaces of almost any shape and texture before drying solid. These were invaluable resources for shipbuilders and barrel-makers and were probably used by the embalmers of ancient Egypt in the hope of making mummies watertight. The discovery that tar could be pressed into service on rough, dusty roads made all the difference to early motorists, too, whose cars ran so much better on Tarmac – though pine tar was soon replaced by more solid, oil-based versions. Pine tar, often golden and relatively fluid, is still good for treating wooden roofs, boats and garden furniture, and used to be applied in efforts to combat dandruff, though with somewhat mixed success. Lucrative pine products from the extensive forests of North Carolina led to its popular name, the Tar Heel State, and baseball teams still use the local product for helping players hang on to the bat handles. Tar can also make a sticky contribution to some kinds of medicinal soap. The thick plentiful resin of these trees has also provided a seemingly inexhaustible resource for glues and waxes, solvents and gums: people tap into the pine for turps, tapes and trips.

The tactile qualities of tar have inspired far more brutal practices, too, including 'tarring and feathering', which involves covering unfortunate victims of collective disapproval in liquid tar and then subjecting them to an avalanche of feathers. Seamus Heaney describes the 'tar-black' face of an ancient body excavated from the Danish peat bogs in his poem 'Punishment', linking the fate of the unknown woman with that of those in modern Belfast, who received similar treatment from the IRA during the Troubles.

The terrors of tar feature a little less grimly in *Great Expectations*, where Dickens turns tar water into an unwanted remedy for young Pip. Tar water, made from diluted pine tar, was one of those traditional cure-alls, to be spooned into children to keep them safe from almost any ailment that might threaten. The philosopher, polymath and highly original thinker George Berkeley, Bishop of Cloyne, claimed that twenty-five fevers in his own family had been cured by tar water, and his best-selling treatise on the subject, *Siris: A Chain of Philosophical Inquiries and Reflexions on the Virtues of Tar Water*, gradually extended into remedial recommendations for the entire country. The flavour that inspired Berkeley and struck horror into the heart of young Pip is still used, albeit more modestly, in the production of liquorice, beer, ice cream and sweets in modern Finland. Antiseptic and fresh-smelling, the pine provides cures for sore throats and bronchitis. In Edwardian England, consumers were urged through contemporary advertisements to buy PEPS – 'A Pine

Air Treatment for Coughs and Colds', which was endorsed by no less a figure than Harry Lauder, who, as everyone knew, needed to take good care of his voice.

Pine needles are still infused into baths for relieving rheumatic aches, but it is a good idea to put the sprays into porous bags for fear of where the needles might end up. This is a tree with an enchantingly heady scent. No wonder bath oils promise the smell of fresh pine – it is all part of our collective pine fantasy. This is the tree that continues to mean 'natural', even when it is most exploited for commercial ends. We like our toilets to smell of the pinewoods – and as pine oil is an important component of many disinfectants, this is not just a question of disguising some of the more natural, if less welcome, odours. Pine pillows, whether infused with manufactured scents or stuffed with fresh needles, help people drift off into clear-headed sleep. You can almost taste the air of a damp, aromatic northern pine forest or a warm, pine-fringed Mediterranean cove, where the perfume seeps down to the sea.

The dark, graceful, floppy-fringed stone pines of southern Europe are also a source of edible pine nuts, which feature in recipes from Italy to the eastern Mediterranean. True pesto originates in the area around Genoa, where the best basil and pine nuts grow. The nuts are pounded with olive oil, pecorino and parmesan to form the oozing sauce for Italian pasta and fish dishes. Roasted pine nuts are sprinkled into salads and savoury dishes in Greece and the Lebanon, while in Tuscany and Turkey they are baked into biscuits, cakes and tarts. You can even make a kind of milk by crushing pine nuts and shaking the mixture up with water, creating a cocktail more nutritious than many. The pine provides wholesome dishes and a natural means of cooking them, scattering huge cones for beach barbecues and aromatic, outdoor fires.

Pine cones have always been a staple of traditional weather forecasting, too, because they relax as temperatures rise, letting the tightly turned, armour-plated fists loosen into layers of open, woody petals. Now meteorologists have realised that their needles are just as revealing. As pesticides and pollutants settle into their green wax-coats, clear patterns are left behind, which, if mapped over months, will give an accurate record of gradual changes in air quality. This absorbent, adaptable tree is a true survivor. After the nuclear disaster at Chernobyl in 1986, some of the pines in Ukraine demonstrated an astonishing capacity to survive the radioactive winter by apparently adapting their DNA to the new toxic environment and so ensuring gradual recovery. The evolutionary implications of this are hard to overstate. Even more exciting potentially is the recent discovery that the lovely scent of pine stimulates the expansion of particles in the air to produce a cooling aerosol effect as they rise. It seems that a pine forest can create its own cloud cover – a huge natural mirror, which will, in turn, reflect some of the sun's rays back into the stratosphere and away from the overheated earth. As we watch the ways of the mysterious pine, we discover that this tree is watching over us.

It is too early to know what the full implications of all this might be, but it seems that the pine's time-old tendency to help and heal is coming to the aid of humanity once again. John Muir, the laureate of the great American forests, was confident that the earth had no sorrow that it could not heal and his own reverence for the teaching of pine trees may be proving very well founded. Going back to nature may be a perennial fantasy, but it seems that we still have a lot to learn from the pine.

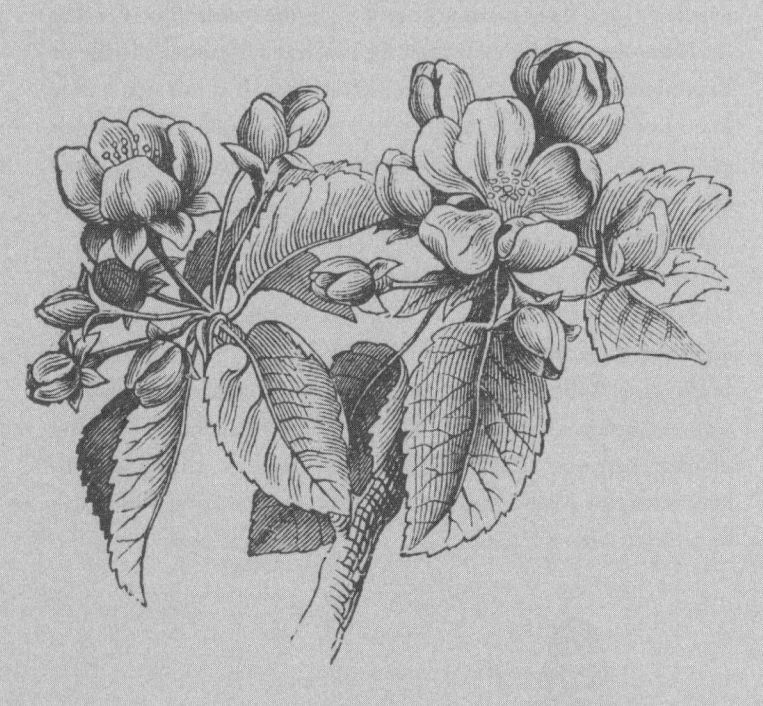

APPLE

THE apple tree is the arboreal alpha – always there at the begin-ning of things. Western culture begins with the apple – whether we go back to the Garden of Eden or to Ancient Greece. In Genesis, the tree is not actually specified, but Renaissance painters and poets were in no doubt about what had seemed so delightful and inviting to our great, great, great grandmother, Eve. Milton imagined the Satanic serpent winding up 'the mossy trunk' of the Tree of Knowledge, 'to satisfy the sharp desire . . . of tasting those fair apples', while in the arresting compositions of Dürer or Cranach, Titian or Rubens, the apple tree stands upright between the first man and woman, hung with smooth, plump, irresistible orbs. But why the apple tree?

If you see a mature tree in September, branches bent earthward with the weight of new fruit, flushed with fresh colour, smooth-skinned, plump and dimpled, peeping from the strong-veined, green leathery leaves like a vibrant décolletage, you can begin to guess. And those low-growing branches make the forbidden fruit so *very* easy to reach. This is the tree of beginnings, but it is also the tree of temptation, the tree that gets the blame when everything goes pear-shaped. In the Song of Solomon, the apple tree is the most desirable tree in the wood, the nest and larder of lovers, whose very breath is apple-scented. For the Greeks, it was the tree of love and discord, because when faced with the difficult task of choosing between three goddesses all regarding the golden apple as theirs by right, Paris decided that it should go to Aphrodite,

LUCAS CRANACH, *ADAM AND EVE*

goddess of love. The revenge of the rejected goddesses, Hera and Athena, rapidly spiralled into an all-consuming conflict, as Paris gained Helen of Troy but lost everything else in the cataclysmic Trojan War.

MARCANTONIO RAIMONDI, *THE JUDGMENT OF PARIS*

The apple tree fosters love and hatred, as you can see from the way its fruit generally grows, with one fat cheek glowing red in the warm breath of the late summer sun, the other, pressed against the rough branch, left pale and green. The apple is both the close-bosom friend of the maturing sun, loading its branches with blessings, and the fruit of the Poison Tree that is capable of growing so quickly in a mind filled with repressed anger and jealousy. Something about the tree with the perfect, palm-sized spheres prompts deep emotions, as we learn very early from the story of Snow White. Under the lovely, shiny red surface, we sometimes find worm-holes, earwigs and a core completely rotten: not every new bite is as sweet as it promises to be.

ISAAC NEWTON IN THE ORCHARD

A falling apple might seem to mark the end of burgeoning beauty: the moment of fullness, when everything ceases to seem promising. Often a falling apple is really the beginning. When Isaac Newton abandoned his studies at Cambridge in 1665, because of the outbreak of the Plague, he found himself back on the family farm in Lincolnshire. The heavy crop of apples in the orchard was part of the regular pattern of the season, but this year he saw it in an entirely new light. Why did apples fall down towards the ground? Why not spin up into the sky or sideways across the field? For the brilliant young mathematician, a peaceful hour under the apple tree was a moment of revelation and revolution. This was the Tree of Knowledge and a fortunate Fall, for suddenly the movement of the whole solar system was visible in the seasonal windfall.

Newton's tree reached a very ripe old age, eventually succumbing to gravity in 1820, but the orchard remains as a living monument to the irrepressibility of the apple tree: one of the branches from the fallen tree grew into the fat-trunked veteran that still yields enormous ruddy-gold apples in the autumn. This old variety, Flower of Kent, runs through the spectrum from green to orange and red as the fruit ripens. A tiny remnant of the wood from the original tree survives in the form of a snuffbox that is displayed like a sacred relic in the manor house, while at the Isaac Newton Shopping Centre in nearby Grantham there is a large clock on which, on the chime of every hour, a red plastic apple strikes a bell, startling a sleeping lion and unsuspecting visitors. (The tree's powers of survival were severely tested during the wholesale restructuring of the centre, but the lion, the apple and the clock are still there.)

The apple's place as the Tree of Knowledge was reinforced when the rainbow silhouette of an apple with a single leaf became the iconic sign of the first personal computers, linking the age of the gigabyte to the great Newtonian revolution in science. It has also been read as a tribute to Alan Turing, the code-breaker and pioneer of computing, whose suicide in 1954 was induced by the intolerable pressures associated with being gay at a time when it was still illegal in the UK. When his body was found, it was lying beside a poisoned apple, half eaten like Snow White's. The logo may have owed something to the youth revolution of the 1960s, too, since Steve Jobs was thirteen when his favourite band launched their Apple record label.

The apple was a symbol of youth culture thanks to the Beatles, who sent up the corporate world with their own 'Apple Corps'. Disc jockeys loved the first record with the Granny Smith centre, because 'Hey Jude' was the longest single ever produced and its running time of over seven minutes gave them time to grab a quick coffee before the nah-nah-nah-nah-nah-nah-nahs finally faded. The song on the B-side of the sliced apple was 'Revolution'. Beatles fans were so ready to free their minds that they flocked into the new psychedelic Apple Boutique in Baker Street, demonstrating the flaw in the business model by failing to pay for most of what they found there.

In the mythic land of eternal youth, everyone lived on apples, at least according to the ancient Celts. This was the fruit that flourished in the island of Avalon, which Tennyson imagined, 'Deep-meadow'd, happy, fair with orchard-lawns', a haven where the dying King Arthur might be healed of his grievous wound. For the Vikings, too, the powerful alpha-male gods depended on the lovely apples of the goddess Idunn to ward off old age and mortality. The apple tree's long-standing association with youth may have something to do with its relatively short lifespan. Unlike the oak or the yew, apple trees often live for no more than thirty years, shooting up and falling before their slow-growing compatriots have really got started. They age surprisingly quickly, being prone to unfortunate complaints such as apple canker and scab. Even a healthy tree has brown, stubby bark, with branches striking odd angles, as if stooping prematurely: all the goodness of the tree seems to go into the flawless, rosy-cheeked fruit. Some trees do live to eighty or a hundred years or more, but once these older, rather bedraggled specimens stop producing fruit and start dropping branches, their days are usually numbered.

Only two veteran apple trees made it into the Tree Council's Jubilee chart of the fifty Great British Trees in 2002: Newton's historic apple tree at Woolsthorpe and the original Bramley apple tree at Southwell in Nottinghamshire. Britain's best-loved cooking apple is not named after Mary Ann Brailsford, the girl who first raised the tree from a seed during the Napoleonic Wars, but after Mr Bramley the butcher, who bought her cottage garden and saw the first green apples swelling on the branches in the year of Queen Victoria's succession. The apple's true potential was quickly spotted by the Merryweathers, who owned the local nursery and who soon established an orchard from cuttings taken from the butcher's tree. The success of the apples was commensurate with their size, and the familiarity of their name today demonstrates the many different ways in which an apple tree can confer immortality. The celebrity of the Bramley apple did not secure the future of the original tree, which fell to the ground some years later, though with new roots and shoots sprouting from the old trunk and a rather more devoted owner, it has enjoyed a considerable revival of fame and fortune.

As apple trees generally flourish and fall so fast, we might assume that they are quick to multiply. In fact, few apple trees are grown from seed because, being heterozygous, the saplings will generally turn out differently from the parent tree. Despite Mary Ann Brailsford's great success, a tree raised from a seed is unlikely to grow into a healthy, fruit-bearing adult tree – as I discovered in a formative horticultural experiment, when, after planting some pips with great enthusiasm, I watched them grow first into pretty little seedlings before changing gradually into stunted, twisted parodies of trees. One did survive into maturity, but it was never the upright, fruitful tree that I had hoped to see. More experienced growers know that apple trees are best produced by cutting scions – or small twigs – from a healthy tree and grafting them onto rootstock. By crossing different varieties, new kinds of apple can be bred and bred again. The remarkable apple tree belonging to Paul Barnett hit the news in 2013 because it displayed 250 different apple varieties, all carefully grafted onto its hospitable branches. The astonishingly bushy canopy, brim-full of bright fruit and little, fluttering pennants labelling each kind, is almost too much for the trunk, so every branch has its own supporting stick, creating a strange, angular, under-shadow of the exuberant tree.

Far from being an unchanging part of British culture, apple trees have been subject to perpetual modifications of one sort or another

– the old ribbed costards that Shakespeare enjoyed had more or less become extinct by the time Richard Cox retired from his brewing business in 1820 to concentrate on apple cultivation at his estate near Slough. As is the fate of most famous apple trees, the original source of Cox's Orange Pippin blew down in 1911, but by then the demand for these delicious dessert apples was such that there were plenty of flourishing descendants. A display of different apple varieties can be quite a family gathering: the Sturmer Pippin was probably the offspring of the Ribston Pippin and a nonpareil.

Apple-growing is a levelling pursuit, flourishing alike in cottage gardens and great estates: the Blenheim Orange, despite its grand name, was first raised by an Oxfordshire labourer. Where else but in an orchard might you find a Grenadier, the Duke of Devonshire, Lord Lambourne, Lord Burghley and the Prince of Wales cheek to cheek with Annie Elizabeth, William Crump, Reverend Wilkes and Granny Smith? (Granny Smith may well seem rather sour in an English orchard, though, because she really needs the Australian sunshine to bring out her sweeter nature.) For all this clever cultivation, there have always been lucky finds, too – Bess Pool is named after an innkeeper's daughter, who discovered the seedling in the woods one day. Apple history is full of hard graft and happenstance – but it usually involves someone spotting the potential in a humble young tree. Apple names are full of hidden stories, too, but beware: Adams Pearmain has nothing to do with the world's very first gardener – it was named after Mr Robert Adams; and if you thought Newton's Wonder was something to do with Sir Isaac, you would be wrong – it was first cultivated in King's Newton in Derbyshire.

So there is really nothing very *natural* about apple trees. Even their mature silhouettes are often man-made. People like to 'train' apple trees, as if, with a bit of patience, they might learn a trick or two. And they do. Under the care of skilful hands, these trees can be encouraged to assume the most surprising shapes – pyramids or wine glasses, chevrons or peacock-tails. An espalier apple tree, with its branches stretched out to form a great green see-saw, takes its name from the 'shoulder', because that is where the oversized limbs need most support. These carefully trained habits are not purely whimsical: they make picking easier, ensure that the fruit grows more evenly and help each apple soak up the sunrays for all-over colour.

Apples have always been a basic, affordable food, whether eaten straight from the tree or baked in pies, puffs, ambers and dumplings. In Britain, those unable to afford an oven of their own used to take apple dishes to the local bakery to be cooked: these had to be carefully marked, of course, to avoid any pudding rage. Even bitter crab-apples, the fruit of the beautiful, native, wild apple tree, are a seasonal bonus, because they can be made into jelly to go into sandwiches or to accompany meat. Memories of coming home from school in late September are sticky in the mind with the smell of jelly-making and my mother boiling ripe crab-apples, before straining the thick, dark, pink, viscose pulp very slowly through a makeshift jelly bag, formed from an old pillow case firmly suspended from the legs of an upturned bentwood chair. Apples are rich in pectin, the vegetarian's favourite source of gelling agent, so crab-apple pulp sets easily into jars of translucent, sunset-coloured jelly. Apple pectin will also help set plums, blackberries and green tomatoes into jams and chutneys.

The old method of ripening tomatoes was to add an apple to a bagful, which was probably effective because apples naturally release a plant hormone, ethylene. Some varieties have the opposite effect on humans, though, and now stem cells from a rare Swiss variety of apple have been used to boost human skin growth and reduce wrinkles. The apple's capacity for unripening humans is also being studied by researchers working on its potential for preventing certain cancers and vascular diseases. An apple a day may well turn out to be a medical prescription for the twenty-first century. The dream of smooth faces and disease-free bodies is a modern version of the land of eternal youth, and the apple still hangs there, inspiring us to try new ways to reach paradise. Who would have guessed when throwing apple peel over the shoulder to discover the initial letter of their

JOHNNY APPLESEED

future spouse that, in time and in rather different circumstances, the peel would also reveal anti-cancer chemicals, triterpenoids?

The healthy, life-sustaining apple may seem quintessentially English, but the trees with the edible fruit only arrived with the Romans, who planted sweet orchards wherever they went. Apples are foundational to American identity, too, and not just New Yorkers. The legendary Johnny Appleseed, planting his nurseries across the eastern states, established the myth of the strong, wholesome, hard-working all-American farmer – the ideal father figure for households brimming with motherhood and apple pie. We now know from tracing the apple genome, however, that the ancestor of all domestic apple trees originated in the Tian Shen mountains, on the borders of China and Kazakhstan. The name of the capital of Kazakhstan, Almaty, means 'city of apples', and the ancient wild orchards that grow on the surrounding slopes may soon be officially recognised as a World Heritage site.

There are now thousands of different kinds of apple tree, marked on world apple maps that register shifting consumer trends. As tastes change, so do commercial fruit crops – the most popular apple in Britain is now the Gala, a New Zealander by origin, very sweet in taste. The recent fashion for exciting fruit-flavoured ciders has also prompted massive tree-planting programmes. We may like to imagine more traditional cider-makers, wheeling their picturesque presses from orchard to orchard and squeezing out the last oozings of summer, in barns heavy with sweat and the heady smell of ripe apples, where apple-pickers are paid in cider, which does little to accelerate the process. Modern fruit-growing is, in contrast, a highly commercial and efficiently mechanised operation. In 2012, international competition led to new EU laws for protecting special regional products, so 'Herefordshire cider' as it is now designated, can only be made from the juices of bitter-sweet apples from local trees such as Brown Snout, Bulmer's Norman, Chisel Jersey or Kingston Black. These varieties are unlikely to be confused with French cider apples, though Bulmer's Norman is a reminder of a shared heritage.

Changing tastes and commercial pressures mean that many of the old, colourfully named varieties have more or less disappeared from the United Kingdom, but efforts are being made to preserve traditional orchards. London may not seem the likeliest venue for apple initiatives, but in October 1990 it hosted the very first Apple Day. For a few heady hours, Covent Garden was reclaimed by Common Ground, the charity committed to recovering links between communities and the natural world, and highlighting regional distinctiveness. Apple Day was intended to raise orchard awareness in an event redolent of Old England and half-forgotten folk calendars. This was a festival for fruit-growers, for people who love glossy pictures of country life and for those who live on health food, dreaming of being close to nature.

Apple Day is the New Age St George's Day, where people from across the political and social spectrum meet annually over apple pie and cider. The numerous regional varieties make apples a perfect symbol of local distinctiveness as well as shared experience, so Apple Days have now bobbed up all over the country, their date depending a little on local varieties and picking times. In mid-Wales there is an Apple Day near Rhayader in mid-September; in the Scottish Borders it falls in early October. Even the woodchips from chopped-down trees come into play for apple-flavoured barbecues. Bacon smoked with apple-wood is a happy addition for those less enthusiastic about all the vegetarian outlets and the high-minded, seed-topped 'apple crumbles'. Such blatant invention of tradition may raise sceptical eyebrows, but also, surely, a certain admiration for the marketing acumen of the apple-growers and their determination to remind modern consumers of the sources of their daily fare.

Children already associate apples with IT, but with the help of the tree's electronic namesake, they can see the apple-pickers of Kashmir or Chile and think about the kind of labour that went into the neat shiny six-pack on the supermarket shelf. An apple offers a direct connection between people on different sides of the globe. Before the end of apartheid in South Africa, Granny Smiths from that country became Pomona non grata in many European kitchens, though there was always a question over whether a fruit boycott was helping or harming those working in the orchards. There is still much to be gleaned from the non-electronic apple, too. An hour under an apple *tree* could sow the seeds for not only farmers and gardeners, but also future botanists, chemists, physicists, artists, writers, economists, politicians or business tycoons.

CLARE LEIGHTON, *SEPTEMBER: APPLE PICKING*

It is much easier to grasp that the fruit in the fridge comes from a real orchard once you have seen them growing. You can tell when an apple is ready to pick by giving it a gentle twist: if it is already ripe, the stalk will just give way, freeing the apple to rest in your hand. Unless harvested at the right moment, apples come plummeting down like cricket balls, bruising as they land. Often they will lie hidden in the long grass, waiting quietly until someone feels that telltale, mushy-brown squish underfoot. These trees offer a way into nature's alphabet – and they speak to all our senses. We can touch the rough bark, smell the ripe fruit, listen to the bees humming or the woodpeckers drumming on dead branches. Old orchards are a haven for some of the rarest and most attractive species, for even though the great-crested newts or noble chafer beetles may be in hiding, the small patch of a brown tree creeper might be moving slowly up and down the trunk. A moonlit evening in September might even reveal the pale humpback of a badger feeding on windfalls, or a fox stretching a long, elegant snout to reach the fruit in the lower branches.

The appeal of an apple tree is not just autumnal, of course. For van Gogh it was the blossom that mattered, because the sudden transformation of the most unlikely, misshapen branches meant the blessing of creativity. Breathtakingly beautiful, that burst of pale, feather-bed petals garlands even the most gnarled old tree with clouds of glory. When at last the dawn begins to creep in a little earlier, you can almost see these trees yawn and stretch and shake their twigs into consciousness. The buds seem a little uncertain at first but, defying the threat of lingering frosts, gradually each bough lights up, like a sleepy smile breaking into an overwhelming YES. Something of this seasonal stirring is there in van Gogh's *Apple Blossom*, where the butterfly blooms are almost dancing with the twisting twigs against the bright turquoise light of the sky.

Camille Pissarro painted the apple trees of Picardy again and again, as awed by their transfiguration in spring as by their summer richness or bare winter silhouettes. The white, cotton chemise of the spring goes just as well with the overall mood as the dappled apricot and pan-tile colours of the warm autumn. Pissarro's apple-centred world is a working pastoral, attentive to aching limbs and heavy barrows, but the real local characters in the fields around Eragny are the highly distinctive fruit trees.

By the time the English war artist Paul Nash arrived in northern France in 1917, the landscape was utterly different. He wrote to his wife from the trenches, describing the remains of a French village, in 'heaps of bricks, toast-rackety roofs and halves of houses here and there among the bright trees and what remains of the orchards'. Nash's striking pictures of the blasted orchards, crowns blown off, trunks outlined like glass shards, turned apple trees into blackened, contorted figures, reaching hopelessly for the sun. His powerful images are among the most eloquent expressions of the sheer horror of the First World War and depend on a deep, collective understanding

PAUL NASH, *CHAOS DECORATIF*, 1917

of what apple trees really mean. This is the tree that should have
gone on growing peacefully in the orchards of France, England and
Germany, where its fruit would have been picked, eaten and drunk
by the latest generation of young men, just as it had been by their
fathers and grandfathers. The apple tree is the Tree of Life, but it is
also the Tree of Knowledge, the source of the taste of good and evil, of
the fruit humans cannot resist and will always seize, unmindful of the

AGNES MILLER PARKER, APPLE BLOSSOM

consequences. The ravaged orchard showed what could not be told by shell-shocked survivors. What new beginnings were possible in this hideous, barren waste?

And yet, there is a deep-down, perennial urge to recover paradise, to start afresh. Dylan Thomas was born in 1914, just after the war broke out, but when he looked back on his formative years, he remembered being 'young and easy under the apple boughs' in an idyllic childhood, 'happy as the heart was long'. No blame attaches to the apple tree, here, which glows in the eternal light of lost youth, but the poet, writing in 1945, knew all about the chains of time as he looked back wistfully on his early reign as Prince of the Apple Towns. *Cider with Rosie*, Laurie Lee's memoir of a Cotswold childhood in the 1920s, takes its title from his first taste of that 'golden fire, juice of those valleys and of that time, wine of wild orchards, of russet summer, of plump red apples, and Rosie's burning cheeks'. It is a vivid personal record of boyhood and the intoxicating apples that grow in paradise. Those born into a post-war world still demonstrate the same old impulse towards Eden, to recovering youth and innocence, to beginning again whatever the odds, whatever the cost, whatever the wreckage all around.

In 1939, Europe was convulsed by war again, but against the sound of guns and air-raid sirens, people went on quietly planting trees and harvesting their fruit. Adrian Bell's *Apple Acre* is a record of a small Suffolk farm in the Blitz, and a testament to the resistant, ever-youthful apple. With the introduction of rationing, home-grown fruit became vital to survival and apples, with their capacity to keep, had a special value. As Britain froze in a limbo of snow and uncertainty, out came the apples from their storage boxes, brighter and redder than when they had been packed away. These shining reminders of high summer were also pledges of better days to come. Apple trees mean beginnings, childhood and the Garden of Eden, but they also mean enlightenment, experience and the future. If the apple has often carried the blame for human mishap, this tree has an unparalleled capacity to go on growing and giving us fresh starts.

ROOTS AND BRANCHES

Addison, Josephine, *The Illustrated Plant Lore* (London: Sidgwick & Jackson, 1985)

Ancient Yew Group, www.ancient-yew.org

Andrews, Malcolm, *Landscape and Western Art* (Oxford: Oxford University Press, 1999)

Anon., *English Forests and Forest Trees* (London: Ingram, Cooke and Co., 1853)

Ballantyne, I. and J. Eastland, *H. M. S. Victory* (London: Leo Cooper, 2005)

Baron, Michael and Derek Denman, *Wordsworth and the Famous Lorton Yew Tree* (Lorton and Derwent Fells Local History Society pamphlet, September 2004)

Barrell, John, *The Idea of Landscape and the Sense of Place* (Cambridge: Cambridge University Press, 1972)

Bate, Jonathan, *The Song of the Earth* (London: Picador, 2000)

Bates, H. E. and Agnes Miller Parker, *Through the Woods*, 2nd edn (London: Gollancz, 1969)

Bean, W. J., *Trees and Shrubs Hardy in the British Isles*, 6th edn, 3 vols (London: John Murray, 1936)

Beer, Gillian, *Darwin's Plots*, rev. edn (Cambridge: Cambridge University Press, 2009)

Bell, Adrian, *Apple Acre* (Wimborne Minster: Little Toller, 2012)

Beresford-Kroeger, Diana, *The Global Forest* (New York and London: Penguin, 2010)

Berger, Stefan, *Germany (Inventing the Nation)* (London: Bloomsbury, 2004)

Bishop, Edward, *The Wooden Wonder*, 3rd edn (Shrewsbury: Airlife Publishing, 1995)

Bonehill, John and Stephen Daniels, *Paul Sandby (1731–1809): Picturing Britain* (London: Royal Academy, 2009)

Breslin, Theresa, *Whispers in the Graveyard* (London: Egmont, 1994)

Breuninger, Scott, 'A Panacea for the Nation: Berkeley's Tar-Water and Irish Domestic Development', *Études Irlandaises* 34.2 (2009), 29–41

Brooks, David, 'Napoleon's Trees', *Kenyon Review* 25 (2003), 20–28

Brownell, Morris R., 'The Iconography of Pope's Villa: Images of Poetic Fame', in G. S. Rousseau and Pat Rogers, eds, *The Enduring Legacy: Alexander Pope Tercentenary Essay* (Cambridge: Cambridge University Press, 1988), 133–50

Burns, Robert, *The Poems and Songs of Robert Burns*, ed. James Kinsley, 3 vols (Oxford: Clarendon Press, 1968)

Busch, Akiko, *The Incidental Steward* (New Haven and London: Yale University Press, 2013)

Campbell-Culver, Maggie, *A Passion for Trees: The Legacy of John Evelyn* (London: Eden Project Books, 2006)

—, *The Origin of Plants* (London: Headline, 2001)

Cannizzaro, Salvatore and Gian Luigi Corinto, 'The Role of Monumental Trees in Defining Local Identity and in Tourism: A Case Study in the Marches Region', *Geoprogress Journal*, vol. 1, no. 1 (2014)

Capote, Truman, *Breakfast at Tiffany's* (New York: Random House, 1958)

Cardinal, Roger, *The Landscape Vision of Paul Nash* (London: Reaktion, 1989)

Carey, Frances, *The Tree: Meaning and Myth* (London: British Museum, 2012)

Carmichael, Alexander, *Carmina Gadelica*, 2nd edn., 6 vols (Edinburgh and London: Oliver & Boyd, 1928–71)

Carson, Ciaran, *The Irish for No* (Oldcastle: Gallery Press, 1987)

Clare, John, *The Natural History Prose Writings*, ed. Margaret Grainger (Oxford: Clarendon Press, 1983)

—, *The Poems of the Middle Period 1822–1837*, ed. Eric Robinson, David Powell and P. M. S. Dawson, 5 vols (Oxford: Clarendon Press, 1996; 1998; 2003)

—, *By Himself*, ed. Eric Robinson and David Powell (Manchester: Carcanet, 1996)

Clifford, Sue and Angela King, *England in Particular* (London: Hodder & Stoughton, 2006)

Coleridge, S. T., *The Complete Poetical Works*, ed. E. H. Coleridge, 2 vols (Oxford: Clarendon Press, 1912)

Common Ground, www.commonground.org.uk

Coombes, Allen J., *The Book of Leaves*, ed. Zsolt Debreczy (London, Sydney, Cape Town, Auckland: New Holland, 2011)

Cornish, Vaughan, *Historic Thorn Trees of the British Isles* (London: Country Life, 1941)

Cowper, William, *The Poems of William Cowper*, ed. John D. Baird and Charles Ryskamp, 3 vols (Oxford: Clarendon Press, 1980–95)

Crichton Smith, Iain, *New Collected Poems*, rev. edn (Manchester: Carcanet, 2011)

Culpeper, Nicholas, *The Complete Herbal* (London, 1653)

Daniels, Stephen, *Humphry Repton* (New Haven and London: Yale University Press, 1999)

Darwin, Charles, *On the Origin of Species*, ed. J. W. Burrow (Harmondsworth: Penguin, 1968)

Darwin, Erasmus, *The Botanic Garden* (London, 1791)

Davidson, Alan, *The Oxford Companion to Food*, 2nd edn (Oxford: Oxford University Press, 2006)

Davidson, Thomas, *Rowan Tree and Red Thread* (Edinburgh and London: Oliver & Boyd, 1949)

Deakin, Roger, *Wildwood: A Journey Through Trees* (London: Hamish Hamilton, 2007)

DEFRA, Independent Panel on Forestry, *Final Report*, 2012, www.defra.gov.uk/forestrypanel/reports

—, *Product Specification*, 'Herefordshire Cider', www.gov.uk/government/uploads/system/uploads/attachment_data/file/317262/pfn-herefordshire-cider.pdf

East, Helen and John Madden, *Spirit of the Forest* (London: Frances Lincoln, 2002)

Edlin, H. L., *Collins Guide to Tree Planting and Cultivation* (London: Collins, 1970)

Ehn, Mikael, Joel A. Thornton, Einhard Kleist et al., 'A Large Source of Low-volatility Secondary Organic Aerosol', *Nature* (2014), 506 (7489): 476

Evelyn, John, *Sylva* (London, 1664)

—, *Sylva*, edited with additional notes by A. Hunter (London, 1776)

Everett, Nigel, *The Woods of Ireland, A History, 700–1800* (Dublin: Four Courts, 2014)

Fauvel, J., R. Flood, M. Shortland and R. Wilson, eds, *Let Newton Be!* (Oxford: Oxford University Press, 1988)

Frank, Anne, *The Diary of Anne Frank*, ed. C. Martin (Harlow: Pearson, 1989)

Frost, Robert, *The Collected Poems of Robert Frost*, ed. Edward Connery Lathem (London: Vintage, 2001)

Fulford, Timothy, 'Cowper, Wordsworth, Clare: The Politics of Trees', *John Clare Society Journal* 14 (1995), 47–59

Gayford, Martin, *A Bigger Message: Conversations with David Hockney* (London: Thames and Hudson, 2011)

Gerard, John, *The Herball or General History of Plants* (1597); *Gerard's Herbal*, ed. Marcus Woodward (London: Senate, 1994)

Gibbs, J., C. Brasier and J. Webber, *Dutch Elm Disease in Great Britain* (Farnham: The Forestry Authority, 1994) www.forestry.gov.uk

Gilpin, William, *Remarks on Forest Scenery*, 2 vols (London, 1791)

Gordon, R. and S. Eddison, *Monet the Gardener* (New York: Universe, 2002)

Grahame, Kenneth, *The Wind in the Willows*, illustrated by E. H. Shepard (London, 1931)

Graham-Dixon, Andrew, *A History of British Art* (London: BBC Books, 1996)

Graves, Robert, *The White Goddess* (London: Faber, 1948)

Gray, Thomas, *The Poems of Gray, Collins and Goldsmith*, ed. Roger Lonsdale (London: Longman, 1969)

Grieve, M., *A Modern Herbal*, ed. C. F. Leyel, rev. edn (London: Cape, 1973)

Grigor, James, *The Eastern Arboretum or Register of Remarkable Trees, Seats, Gardens, etc., in the County of Norfolk* (London, 1841)

Groom, Nick, *The Seasons* (London: Atlanta, 2013)

Guyatt, Mary, 'Better Legs: Artificial Limbs for British Veterans of the First World War', *Journal of Design History* 14, no.4 (2001), 307–25

Hadfield, Miles, *British Trees* (London: Dent, 1957)

Hageneder, Fred, *Yew: A History*, rev. edn (Stroud: The History Press, 2011)

Hardy, Thomas, *The Woodlanders* (London, 1887)

—, *Tess of the D'Urbervilles* (London, 1891)

Hawkins, C., *Rowan: Tree of Protection* (privately printed, 1996)

Haycock, David Boyd, *A Crisis of Brilliance* (London: Old Street Publishing, 2009)

Heaney, Seamus, *Death of a Naturalist* (London: Faber, 1969)

—, *North* (London: Faber, 1975)

—, *Field Work* (London: Faber, 1979)

—, *Preoccupations: Selected Prose 1968–78* (London: Faber, 1980)

—, *Sweeney Astray* (London: Faber, 1984)

—, *Opened Ground: Poems 1966–1996* (London: Faber, 2000)

Hemery, Gabriel and Sarah Simblet, *The New Sylva* (London: Bloomsbury, 2014)

Hight, Julian, *Britain's Tree Story* (London: National Trust Books, 2011)

Hoadley, R. Bruce, *Understanding Wood: A Craftsman's Guide to Wood Technology*, 2nd edn (Newtown: Taunton Press, 2000)

Hobson, D.D., 'Populus Nigra L. in Ireland: An Indigenous Species?', *Irish Naturalist's Journal* 24, no.6 (1993), 244–7

Hockney, David, *A Bigger Picture* (London: Royal Academy of Arts, 2012)

Hopkins, G. M., *The Poems of Gerard Manley Hopkins*, ed. W. H. Gardner and N. Mackenzie, 4th edn (Oxford: Oxford University Press, 1970)

Hoskins, W. G., *The Making of the English Landscape* (London: Hodder & Stoughton, 1955)

Housman, A. E., *The Collected Poems*, rev. edn (London: Cape, 1960)

Hughes, Ted, *Collected Poems* (London: Faber, 2003)

Hur, M., Y. Kim, H.-R. Song, J. M. Kim, Y. I. Choia and H. Yin, 'Effect of Genetically Modified Poplars on Soil Microbial Communities during the Phytoremediation of Waste Mine Tailings', *Applied and Environmental Microbiology* (November 2011) vol. 77, no. 217611–19

Jackson, Kenneth, *A Celtic Miscellany*, rev. edn (Harmondsworth: Penguin, 1971)

Jamie, Kathleen, *The Tree House* (London: Picador, 2004)

Jellicoe, G., S. Jellicoe, Patrick Goode and Michael Lancaster, *The Oxford Companion to Gardens* (Oxford: Oxford University Press, 1986)

Jenkins, David Fraser, *Paul Nash: The Elements* (New York: Scala Press, 2010)

Johnson, Hugh, *Trees*, rev. edn (London: Mitchell Beazley, 2010)

Jones, Gareth Lovett and Richard Mabey, *The Wildwood: In Search of Britain's Ancient Forests* (London: Aurum, 1983)

Juniper, Barrie and David Mabberley, *The Story of the Apple* (Portland: Timber Press, 2009)

Keats, John, *The Complete Poems*, ed. Miriam Allott (London and New York: Longman, 1970)

Kilvert, Francis, *Kilvert's Diary 1870–1879*, ed. William Plomer (London: Cape, 1944)

Knight, Richard Payne, *The Landscape*, 3 vols (London, 1794)

Lawrence, D. H., *Sons and Lovers*, ed. David Trotter (Oxford: Oxford University Press, 1995)

Lippard, Lucy, *The Lure of the Local*, rev. edn (New York: The New Press, 2007)

Lee, Laurie, *Cider with Rosie* (London: Hogarth Press, 1959)

Leslie, C. R., *Memoirs of the Life of John Constable*, 2nd edn (Oxford: Phaidon, 1980)

Long, Richard, *Heaven and Earth* (London: Tate, 2009)

Longley, Michael, *Collected Poems* (London: Jonathan Cape, 2007)

Loudon, J. C., *Observations on the Formation and Management of Useful and Ornamental Plantations* (Edinburgh, 1804)

—, *Arboretum et fruticetum britannicum*, 8 vols (London, 1838)

—, *In Search of English Gardens: The Travels of John Claudius Loudon and his Wife, Jane*, ed. Priscilla Boniface (London: Lennard Books, 1988)

Mabey, Richard, *Plants with a Purpose* (London: Collins, 1977)

—, *Gilbert White* (London: Century Hutchinson, 1986)

—, *Flora Britannica* (London: Sinclair-Stevenson, 1996)

—, *Nature Cure* (London: Pimlico, 2006)

—, *Beechcombings* (London: Vintage, 2008)

—, *Weeds* (London: Profile, 2010)

McAllister, H., *The Genus Sorbus: Mountain Ash and Other Rowans* (London: Kew, 2005)

McCracken, David, *Wordsworth and the Lake District* (Oxford: Oxford University Press, 1985)

Macfarlane, Robert and Dan Richards, with illustrations by Stanley Donwood, *Holloway* (London: Faber, 2013)

Maclean, Sorley, *Collected Poems, White Leaping Flame* (Edinburgh: Birlinn, 2011)

McNeill, F. Marian, *The Silver Bough* (Edinburgh: Canongate, 1989)

McNeillie, Andrew, *Now, Then* (Manchester: Carcanet, 2002)

Macpherson, James, *The Poems of Ossian*, ed. Howard Gaskill (Edinburgh: Edinburgh University Press, 1996)

Mahood, Molly, *The Poet as Botanist* (Cambridge: Cambridge University Press, 2008)

Manuel, F. E., *A Portrait of Isaac Newton* (Cambridge: Harvard University Press, 1968)

Marvell, Andrew, *The Poems of Andrew Marvell*, ed. Nigel Smith (London: Routledge, 2006)

Mason, Laura, *Pine* (London: Reaktion, 2013)

Miles, Archie, *A Walk in the Woods* (London: Frances Lincoln, 2009)

Mills, A. D., *A Dictionary of British Place Names*, rev. edn (Oxford: Oxford University Press, 2011)

Milne, A. A., *Now We Are Six* (London: Methuen, 1927)

Milne-Redhead, E. W. B. H., 'The B. S. B. I. National Poplar Survey', *Watsonia* 18 (1990), 1–5

Milton, John, *The Complete Poems*, ed. J. Carey and A. Fowler, 2nd edn (London: Longman, 1998)

Mitchison, Rosemary, *Agricultural Sir John. The Life of Sir John Sinclair of Ulbster. 1754–1835* (London: Geoffrey Bles, 1962)

Morton Parish Council, 'Trees of Morton', http://mortonparishcouncil.org.uk/ history/trees-of-morton/golden-jubilee-horse-chestnut

Mountford, Charles, *The Dreamtime* (Rigby: Gumtree, 1965)

Muir, John, 'The National Parks and Forest Reservations', Proceedings of the Meeting of the Sierra Club, 23 November 1895, *Sierra Club Bulletin* (1896)

—, *Journeys in the Wilderness: A John Muir Reader*, ed. G. White (Edinburgh: Birlinn, 2009)

Nash, Paul, *Outline. An Autobiography and Other Writings* (London: Faber, 1949)

National Hedgelaying Society, *35th National Hedge Laying Championships, Waddesdon Estate* (2013)

National Trust, *Powis Castle* (National Trust, 1987)

Natural England, 'Traditional Orchards: Orchards and Wildlife', 2nd edn (October 2010), n. TIN020, www.naturalengland.org.uk

Newing, F. E., R. Bowood and R. Lampitt, *Plants and How they Grow. A Ladybird Book* (Loughborough: Wills and Hepworth, nd)

Nicholson, B. E. and A. R. Clapham, *The Oxford Book of Trees* (London: Oxford University Press, 1975)

Norton, Peter, 'The Lost Yew of Selborne' (2013), www.ancient-yew.org/userfiles/ file/selborne.pdf

Oak Processionary Moth (Thaumetopoea processionea), www.forestry.gov.uk/ oakprocessionarymoth (accessed 29 March 2016)

O'Brien, David, 'Antonio Canova's *Napoleon as Mars the Peacemaker* and the Limits of Imperial Portraiture', *French History* 18, no.4 (2004), 354–78

Orchard Network, 'Apple Day', www.orchardnetwork.org.uk

Oswald, Alice, *Woods etc.* (London: Faber, 2005)

Pakenham, Thomas, *Meetings with Remarkable Trees* (London: Weidenfeld & Nicolson, 1996)

Paterson, Leonie, *How the Garden Grew: A Photographic History of Horticulture at the Royal Botanic Garden Edinburgh* (Edinburgh: Royal Botanic Garden History, 2013)

Plath, Sylvia, *Ariel* (London: Faber, 1965)

Pliny, *Natural History*, 10 vols, vol. 4, libri xii-xvi, trans. H. Rackham, rev. edn (London and Cambridge: Heinemann and Harvard University Press, 1968)

Press, Bob and David Hosking, *Trees of Britain and Europe* (London: New Holland, 1992)

Price, Uvedale, *An Essay on the Picturesque* (London, 1794)

Rackham, Oliver, *Trees and Woodland in the British Landscape*, rev. edn (London: Dent, 1990)

—, *Ancient Woodland: Its History, Vegetation and Uses in England*, rev. edn (Castlepoint Press, 2003).

—, *The Ash Tree* (Toller Fratrum: Little Toller, 2014)

Rice, Matthew, *Village Buildings of Britain* (London: Little, Brown, 1991)

Richens, R. H., *The Elm* (Cambridge: Cambridge University Press, 1983)

Robinson, John Martin, *Felling the Ancient Oaks* (London: Aurum, 2011)

Robinson, Phil, *Under the Punkah* (London: Sampson Low, 1881)

Roebuck, P. and B. S. Rushton, *The Millennium Arboretum* (Coleraine: University of Ulster, 2002)

Rosenthal, Michael, *British Landscape Painting* (Oxford: Phaidon, 1982)

— and Michael Myrone eds, *Gainsborough* (London: Tate, 2002)

Sanders, Rosanne, *The English Apple* (London: Phaidon, 1988)

Scottish Literary Tour Company, *Land Lines: An Illustrated Journey through the Landscape and Literature of Scotland* (Edinburgh: Polygon, 2001)

Shelley, P. B., *Shelley's Poetry and Prose*, ed. D. Reiman and S. Powers (New York: Norton, 1977)

Shenstone, William, *The Works in Verse and Prose, of William Shenstone, Esq.*, 2 vols (Edinburgh, 1765)

Sheppard, J., *On Trees and their Uses* (London, 1848)

Smit, Tim, *The Lost Gardens of Heligan* (London: Gollancz, 1997)

Soar, Hugh D. H., *The Crooked Stick: A History of the Longbow* (Yardley: Westholme, 2009)

Squire, Charles, *The Mythology of the British Islands* (London: Gresham, 1905)

Stafford, Fiona, *The Last of the Race: The Growth of a Myth from Milton to Darwin* (Oxford: Clarendon Press, 1994)

—, *Starting Lines in Scottish, Irish and English Poetry from Burns to Heaney* (Oxford: Oxford University Press, 2000)

—, *Local Attachments* (Oxford: Oxford University Press, 2010)

Straw, Nigel and Alice Holt, 'Insects of Oak and their Role in Oak Decline', Forestry Research www.forestry.gov.uk/pdf/fhd_22Jun10_insects_of_oak. Accessed 29 March 2016

Strutt, Jacob, *Sylva Britannica: Or Portraits of Forest Trees* (1822), enlarged edn (London, 1830)

Summerfield, Chantel, 'Trees as Living Museum: Arborglyphs and Conflict on Salisbury Plain', in Nicholas Saunders, ed., *Beyond the Dead Horizon: Studies in Modern Conflict Archaeology* (Oxford: Oxbow Books, 2012), 159–71

Sweetman, Bill and Rikyu Watanabe, *Mosquito* (London: Jane's Publishing, 1981)

Syme, Alison, *Willow* (London: Reaktion, 2014)

Taplin, Kim, *Tongues in Trees: Studies in Literature and Ecology* (Bideford: Green Books, 1989)

Tayler, M. and D. Harris, *Batsford Arboretum* (Box-3, nd)

Tennyson, Alfred, *The Poems*, ed. Christopher Ricks, 2nd edn (London: Longman, 1969)

Thomas, Dylan, *The Poems*, ed. Daniel Jones (London: Dent, 1971)

Thomas, Edward, *Collected Poems* (London: Faber, 1920)

—, *A Literary Pilgrim in England* (Oxford: Oxford University Press, 1980)

Thomas, Keith, *Man and the Natural World* (London: Allen Lane, 1983)

Train, John, *The Olive Tree of Civilisation* (London: ACC Art Books, 1999)

Tree Council, www.treecouncil.org.uk

Tudge, Colin, *The Secret Life of Trees* (London: Allen Lane, 2005)

Turner, Roger, *Capability Brown and the Eighteenth-century English Landscape*, 2nd edn (Chichester: Phillimore, 1999)

Virgil, *Eclogues, Georgics, The Aeneid*, tr. H. R. Fairclough, 2 vols, rev. edn (Cambridge and London: Harvard University Press, 1923)

Warren, Piers, *British Native Trees: Their Past and Present Uses* (Wildeye, 2006)

Watkins, Charles, *Trees, Woods and Forests* (London: Reaktion, 2014)

— and Ben Cowell, *Uvedale Price (1747–1829): Decoding the Picturesque* (Woodbridge: Boydell, 2012)

Wells, Diana, *Lives of the Trees* (Chapel Hill: Algonquin, 2010)

Westwood, Jennifer and Jacqueline Simpson, *The Lore of the Land* (London: Penguin, 2005)

Whelan, Kevin, *The Tree of Liberty* (Cork: Cork University Press, 1996)

White, Gilbert, *The Natural History of Selborne*, ed. Anne Secord (Oxford: Oxford University Press, 2013)

White, John, *Black Poplar: The Most Endangered Native Timber Tree in Britain* (Farnham: Forestry Authority, 1993) www.forestry.gov.uk

Wilcox, Timothy, *Francis Towne* (London: Tate, 1997)

Withering, William, *Arrangement of British Plants*, 3rd edn, 4 vols (London, 1796)

Wood Database, www.wood-database.com

Woodland Trust, *Broadleaf*, 2005–15, www.woodlandtrust.org.uk

Wordsworth, Dorothy, *The Journals of Dorothy Wordsworth*, ed. Ernest de Selincourt, 2 vols (Oxford: Clarendon Press, 1941)

Wordsworth, William and S. T. Coleridge, *Lyrical Ballads*, ed. Fiona Stafford (Oxford: Oxford University Press, 2013)

—, *Guide to the Lakes*, ed. Ernest de Selincourt, Preface Stephen Gill (London: Frances Lincoln, 2004)

—, *William Wordsworth: Twenty-first Century Oxford Authors*, ed. Stephen Gill (Oxford: Oxford University Press, 2011)

Wullschleger, Stan D., D. J. Weston, S. P. DiFazio and G. A. Tuskan, 'Revisiting the Sequencing of the First Tree Genome: *Populus trichocarpa*', *Tree Physiology* (2013), 4–23

Yeats, W. B., *The Poems*, ed. Daniel Albright, rev. edn (London: Dent, 1994)

Young, John, *Robert Burns: A Man for All Seasons* (Aberdeen: Scottish Cultural Press, 1996)

INDEX

Illustration Credits

Agnes Miller Parker illustrations from H. E. Bates, *Through the Woods*, 1936, 1969: 3. Acorns, Berries, Chestnuts, Hazelnuts; 96. Oak Buds; 208. Willow Catkins; 241. Red Squirrel among Pine Cones; 270. Apple Blossom.

Alamy: 6, 7, 13, 40, 66, 80, 82, 101, 102, 105, 118, 132, 216, 233, 258.

Gwen Raverat (reproduced by permission of The Gwen Raverat Archive, www. raverat.com): 130. *Poplars in France*; 180. *Horse Chestnuts at Grantchester*.

28. Geograph, Julian P. Guffogg. 54. Geograph, Claire Pegrum 86. Photographer A. D. Richardson, 17 July 1894, Royal Botanic Garden Edinburgh. 186. Getty Images. 189. Malcolm Sparkes, *Horse Chestnut Leaves* (reproduced by kind permission of the artist). 199. William Henry Fox Talbot, *Elm in Winter*, ca. 1845, salted paper print: 7 x 61/4 in. (17.78 x 15.88cm), San Francisco Museum of Modern Art, Accessions Committee Fund purchase. 210. Pope's villa at Twickenham (reproduced by kind permission of the Richmond Borough Art Collection, Orleans House Gallery). 214. E. H. Shepard illustration from Kenneth Grahame, *The Wind in the Willows*, 1931 (reproduced by permission of Curtis Brown Ltd). 230. *Look and Learn*. 24, 267. Clare Leighton, *September: Apple Picking*, Boston Public Library catalogue no. 209, Bridgeman Art Library (reproduced courtesy of the artist's estate). 288. Illustration attributed to James McNab, son of William McNab, who masterminded the 1820 relocation of the Royal Botanic Garden, Edinburgh, from Leith Walk to Inverleith (reproduced by permission of Royal Botanic Garden, Edinburgh).

Every reasonable effort has been made to trace copyright holders and obtain their permission for use of licensed material. The publisher apologises for any omissions in the above list and would be grateful for notification of any copyright information that should be incorporated in future reprints or editions of this book.

BC 25/17
